BIGFOOT
Methodology of Madness
By, Bea Mills

ISBN: 978-1-940087-56-6
Publisher: 21 Crows Dusk to Dawn Publishing,
21 Crows, LLC

 This is a true story based on actual events. This book was printed in the United States of America.

Table of Contents

Forward

If you are reading this, then you must be mad. Not in the traditional sense of mad, more like the Hatter from Alice in Wonderland, mad.

Who are we that traverse the countrysides, meadows, creeks, mountains, and everywhere in between to look for a two-legged entity we call by many names? Spending ridiculous amounts of money, time, and resources while enduring dubitable relationships and drama. And yet, we make new friends from all over the world and always have a new place to explore. Deliberating for hours, even days, by every means of communication, to collectively agonize over any piece of evidence brought forward that leads to even stranger conversations. Facing reactions of indifference, acceptance, or resentment that typically boil down to confusion, shrugged shoulders, and eventual withdrawal. Both men and women have devoted entire portions of their lives to solving this case. And many more have left this worldly existence with nothing to satisfy their thirst for knowledge or our own morbid curiosity.

Oh, what is this relic man, beast, ape, alien, portal hopping mystery whom we so lovingly call "Bigfoot"? Whether diving into handed-down legends and lore or spending time on modern-day social media - there are no clear answers to all the questions. Results in evidence have continued to lead us nowhere, yet despite the madness, we continue to press on. Still exploring, theorizing, sometimes fantasizing, but always hoping for even just a glimpse of this enigma, Bigfoot.

We must be mad.

Chapter 1

Bigfooting is every puzzle, sport, or game you have ever played that has a slim to none chance of ever winning. Yet, here we are, NOWHERE. – Bea Mills

—Nowhere—

Collections of stories passed down through the generations have built the framework for what, or who, we refer to as the modern-day Sasquatch. Many of the 'classic' ones were adopted from news articles. *BIG FOOT* came from an interview with Jerry Crew back in 1958. The *GRASSMAN* was born from an article in Akron, Ohio. The more you look, the more you will find that many rural localities have their own name to share. Just a few are the Jim Hinker, Patty, Clear Creek Monster, MoMo, Mr. Howdy, Harry, Big Moe, and the Minerva Monster.

Each name often has its own collection of stories. Up until recently, this field was, and still can be, taboo. Getting folks to share their experiences can be quite challenging. Still, thankfully, there are a few truly dedicated investigators and historians who have done both field time and their homework to preserve these stories for us.

Their documentaries and books have provided a historical treasure trove of information on the enigma. Besides, it's always fun finding out, how'd that Bigfoot get their name?

So, what exactly is a Bigfoot? Take your pick of choices, including Relic Hominid, Alien, Gigantopithecus Blacki, Nephelium, or Transcendent. I will quote Cliff Barackman on this as I found his answer simply perfect: "A Sasquatch is just that, a Sasquatch." In my opinion, it is the only right answer until any hard scientific data becomes available to determine otherwise. No amount of evidence has been quantifiably produced to even acknowledge its existence in any peer-reviewed, academic, scientific, or public respect. Therefore Bigfoot will continue to be a belief, a story, a myth, and its very own legend.

Then, how does one explain the stories that have compelled so many to dive into this world of madness? Is the Bigfoot phenomenon just a trick of the mind? Or does a big, hairy, loud, foul-smelling, two-legged creature that we call Bigfoot really exist? Existential conversations in general life can go a multitude of directions. Now, tuck it directly into the Bigfoot world, and viola, madness! Ask any person from one of the following four groups; the 'believers,' the 'skeptics,' the 'no ways,' and the 'knowers'; each one will give you their opinion, and they are all unique. Sometimes their answers are thoughtful and polite. Other times not so much. Cyclical debates may last for years. The hoaxers, thieves, trolls, and storytellers just pollute and further dilute everyone's slim hopes of truth. Without integrity, a collaboration of cohesive thoughts and ideas, and politely agreeing to disagree, the entire subject becomes lost. This perfect storm can lead any good-intentioned 'Bigfooter' into an abyss of disappointment. These debates on existence lead us nowhere. There simply isn't enough supporting data, either way, worth arguing about.

And yet, in their defense, each group is important. Collectively, they maintain a (very) rocky balance in the Bigfoot world. Does this have anything to do with helping prove the existence of a Bigfoot? Yes, and no. Whatever your personal motivation may be for stepping into this field and, no matter what your existence beliefs are, you will eventually meet these groups. It is in this sense of balance that they force a higher bar. They demand excellence in evidence collection, assimilation, and distribution. And why not? I applaud them all. Without these four respective groups, despite the in-fighting and wars, each one keeps very high expectations from the other.

That said, without hard evidence to show, every enthusiast has an obligation to note that: any report, including their own, is just a story. I am an investigator and have just as high expectations from myself and my own investigators. As such, we all have the duty of due diligence. Opinions are just that—one person's view of the world. No matter which belief you hold to be true, there is still not enough evidence that has been brought forth to prove who is right yet. When the day is done, we all want the same thing —the truth.

Throughout the remainder of this book, I will discuss my own 'citizen scientist' methods, compiled after years' worth of field time. I've included a few of my own experiences, adventures, reports, and research projects. For years, I've been asked about my 'methods.' Where did I learn them? What do I do with the information? And why haven't I written this all out before? Perhaps, I just needed to be properly motivated and have foot surgery to slow me down for a bit.

Having spent most of my life immersed in nature, I've seen and experienced some pretty strange things outdoors. But that is why we call life a journey, right? Reflecting on time spent in the field and around other folks, I realized I have my own investigative style. While I'm usually met with stares and quizzical faces, my methods seem to get the job done with a minimal margin of error. Some call it madness; I call them *The 4 Principles.* It wasn't until formally reviewing my Bigfoot investigation, spanning six years, that it became so plain and obvious an outline. Looking back since early childhood, I realized this has been my 'way' in nearly every endeavor. While the components have matured and changed depending on the situation, the principles have remained unchanged. I utilize these in not only my investigations and research, but even my personal life. They are: *Homework, Strategy and Execution, and Resolution.* Each principle, consists of several individual components. You will find that the majority of the components overlap because they are all *dynamic.* Different aspects need to be examined differently in relation to the individual principles.

At a minimum, each one will be reviewed during the process.

My methodology has helped answer many questions as a field investigator, and although it is ever-evolving, I give it credit to the many successes and learning opportunities, particularly during that long-term investigation. While my audio evidence emerged publicly in 2017, it wasn't until 2019 that all records were broken. An article written in the *Bigfoot Times*, reviewing my work in documenting, casting, and preserving a historical track line find, has been my only official publication. A site interview was conducted for a popular, tenured Bigfoot television show in late 2020. From that trackline, I was honored as the recipient of the "*Bigfoot Times*, 2019 Bigfooter of the Year." First Female.

Was my goal to prove the existence of Bigfoot? Honestly, in 2019 I thought I was close. Close enough to inquire for information on how to put together a journal article for peer review, pending the DNA results, and wanted it done immediately. I had everything but a body. But then, my in-depth trackline report was denied being published, and a few months later, I learned my entire investigation was almost stolen. So now, nope. I am perfectly content doing my thing, having fun, exploring, and being a naturalist, educator, nerd, tree hugger, and mom. You can count that I'm still in the field. With so many questions needing answers, the 4 principles continue to guide me through both field work and everyday life.

So, do you still want to turn the page and give it a go in this crypto Bigfoot world of madness? Do you still want to try and find that elusive mystery of a hairy monster? Then, by all means, DO IT! While you enjoy this maddening hobby we lovingly call Bigfoot, I welcome you to inspect, dissect, and elaborate on my four-step process for yourself. Is it perfect? Absolutely not, but it works for me.

There are many paths and options involved to thoroughly exploring this hobby in that respect. What *you* do is completely up to you, and worst case—just donate this book to the thrift store. But, if you are dedicated to learning and kinda like Bigfoot, I can almost guarantee that you will have fun and learn plenty! Throughout life, people, places, and circumstances change. In turn, we learn, adapt, and overcome. Good humor, a good attitude, and a very healthy dose of skepticism will see you through most things. I applaud your efforts already! I sincerely hope that my work, knowledge, and years' worth of trials and errors may be found useful along your way. Enjoy, and happy Bigfooting!

The best place to start any journey is at the beginning. Let's get out of the middle of NOWHERE.

Chapter 2

"Knowledge for the Sake of Knowledge"
– Dick Geyer, first day of freshman history class written on the chalkboard. 1994.

PRINCIPLE #1: Homework

A) Pieces

B) Parts

Let's start off with the classic—"Yeah, I know." But do you? Do any of us? I am confident that the answer is simply NO. There is always something new to learn and try. As with any journey, when you experience doubt or run into a problem, the best course of action is to collect your thoughts and circle back to the point of deviation. Even if going back means going all the way back to the basics. Just do it. Learn what went wrong, and try again.

This Principle requires the most amount of time, as there are several components involved. Many might seem boring, and even more—ridiculous. Yet, this is the most important one to really spend time working through. Whether you've never been in the field, don't believe in Bigfoot, are a very tenured investigator, or grew up with the Sasquatches themselves, you will still find yourself revisiting this principle, HOMEWORK, quite often. So, I broke this into two big categories—*pieces* and *parts*. Each one has sub-topics to explore. This is by no means dogma; I simply want to get you thinking. The odds are brilliant that you will come up with additions of your own and name them strange things too. For me, I have *pieces* and *parts*. *Pieces* are snips of relatable information, and *Parts* make them relevant.

Both are nothing more than one tiny thing, but that one tiny thing is a specific answer. The answers give me details, and each detail contributes to the bigger picture. This concept is basically our everyday life. We must look at the details when we *want*, *start*, or *try* something new.

Example: I want a diesel truck. Where do I buy a truck? New or used? Ford or Chevy? Black or white? Lifted? Turbo? Which model year was best? Which engine do I pick? What are the recent reviews? What is a fair price? Why are eggs five dollars in Ohio right now? How much is diesel at the pump today? Is four-wheel drive an option? What's the towing capacity? Now I need a camper! What size rims do I want? Trim packages and neon? What sticker should I put on first? Blah. Blah. Blah. Blah.

If you didn't know much about buying a diesel truck, then that is probably overwhelming. It is maybe enough to make you quit looking before you even start mulling over details. So, let's remove that feeling. After all, this is Bigfoot, and Bigfoot is supposed to be fun. You already know what pieces and parts are in the real world, so let's adjust your perspective slightly because we're not looking for a truck; we're looking for the world's hide-and-seek champion. The fact that no one can prove it exists makes it more fun because you can't be sad you didn't find him!

Early in my Squatching days, I had no idea where to start getting 'in the field.' I didn't even know what that meant in Bigfoot terms, so I literally just drove around to parks and forests to hike in the middle of - nowhere. I spent a good year wandering. The sightseeing was amazing, and the places I found were beautiful, but I noticed how the land always seemed to be changing. Some forests were gorgeous and easy to walk through; others were death marches through acres of prickers, barbs, and nasties that tore up my skin and clothes.

I saw streams and run-offs that were rust orange and smelled horrible, while others were crisp and clear, teeming with fish and crawdads. While it was fun exploring, I just figured that getting bloody was part of this thing called Squatching. Well, it's not. It was because I didn't know much. I was in all the wrong areas of the right regions. But, once I learned a few things, and with good old common sense, Squatching got much more interesting and a lot more fun. That progression of learning hasn't stopped since.

Television. That vacuous screen of mind-numbing images and noise. Nearly everyone has watched a show or even taken part in something called a 'case' -an event that happened where there are questions. You may have noticed that the people investigating the case will ask the witnesses many seemingly, nauseating questions. The questions can range from general to specific and sometimes just completely off the wall.

Nevertheless, that investigator is gathering slices of information to create or re-create the history of the case. And that is the key word - History. Let's be honest. By the time you get around to documenting, investigating, or telling any manner of a Bigfoot (or life) story, it's already history. The leaves may have changed, the crops harvested, the snow fallen, the wildflowers bloomed, the rain came, the grand conjunction of planets passed, and a whole new tv series started. The list is endless, but knowing history will directly influence the course of your decisions for the long haul in the crypto world. So, let's get some knowledge for the sake of knowledge and start at every investigation's beginning- history. So what? History's dumb. Let's go Squatching, storm the woods, and find that Bigfoot! Where are we going? Easy, we go outside. That doesn't help! I want somewhere good! I'd hope so. First, do your homework: Learn your Land. Read a Map. Make a Map.

PIECES

Land and History

The beginning starts with the land. Your homework is to find pieces of relatable information that look like 'somewhere good' to begin your new journey in Bigfoot. Land has been around since Pangea and is always moving on those tectonic plates. (lose you, yet?) Over the last few centuries, we have contributed to innumerable changes in our natural areas.

Clearcut forest after with newer growth.

Old Hope Furnace—Lake Hope State Park.

Some were good, some were bad, and many were brutal. Two centuries ago, railroads and canals were built. Entire forests were clear-cut, and swamps drained. Old towns and homes were flooded to control the water and make lakes and reservoirs. Strip mining, coal mining, and blasting hillsides and mountains were normal. We are constantly building new freeways for travel and roads for urban sprawl. Pipelines, powerlines, and oil and gas wells look like their own messy highways through nature.

Think back to where you grew up. I'm sure some buildings changed in just a few years, homes were built or torn down, and shopping or gas stations might have opened nearby. That old field you used to walk home from school or look for arrowheads might now be a subdivision, school, or industrial park. Even though there have been many great strides to conserve, preserve and downright save our greenspaces, the sad truth is that we will never see them restored to their former glory. Just like the plants and animals, we are forced to adapt to new, changing environments. We have to learn new ways to navigate through them. What may have been a viable habitat for Sasquatch one, five, ten or one-hundred years ago might not be suitable today. History provides us with clues for the big picture.

In terms of reports, Ohio's documented reports of Bigfoot started in the late 1800s and all but disappeared in most regions until the late 1950s. Why? History. I see a heavily forested area that looks like it might be 'somewhere good' but totally isn't. Why? History. When cross-referencing reports and seeing something interesting that shows up. Why? History. Land's history will usually hand you a decent answer or, at the very least, give you enough to work with to move forward.

The following is an abstract I wrote for my region's history. Since I live and focus in South and Eastern Ohio, I felt I needed to remember these things. I often use this in presentations and circle back as needed. Yours may look very different depending on where you live, or maybe you're lucky to live so remotely that not much has changed that would affect a possible habitat. Efforts during the industrial revolution in the late 1800 and early 1900s severely affected the South and Eastern portions of Ohio through pig iron forging, coal mining, strip mining, and brick factories.

In 1898, there were 1,155 underground coal mines. With no EPA during that time, mine run-off and acid-producing wastes contaminated many local waterways and grounds. Later aided by inventions like the Muskie bucket, Ohio's forests and natural resources took a heavy blow. What wasn't fueled by timber was clear-cut for agriculture and expansion. Ohio, in all her native splendor, boasted approximately 85% forestland yet barely scraped 10% by mid-1930. Between 1933 and 1942, the Civilian Conservation Corps, and Roosevelt's Tree Army, planted 3 billion trees across the entire United States.

CCC working at Zaleski State Forest planting trees in 1936.

Remnants of the coal mining/railway years—Moonville Tunnel.

Yet, by then, much of the native fauna (forget the flora) had already been extirpated. Even the whitetail deer population was scarce until the mid-1970s. Many of these pieces have become historical markers of the region.

Muskie Bucket—Man posing as dirt clod—

They are found scattered in ruins and lost towns, old homesites, and roads, like Moonville. Miner's Park has (just the bucket of the Muskie) you can climb in. Hope Furnace, in Zaleski State Park, depicts just one of the many raw pig iron implements used in the 1800s. Beneath dammed lakes now used for fishing, boating, and recreation are long-gone villages. The Belgian Coke Ovens, among many others, litter the region just waiting for the brave soul to find and explore.

The point is that change is inevitable. It's that one constant in life we can all count on. Fortunately, there are precious few things that will never change; the sun will rise in the east, the north star will be the north star, and history is history. Taking the time to learn land history can be slightly more challenging, although it will benefit you in all sorts of nerdy, weird ways throughout life and along your quest for Bigfoot. Bonus, they make great road trip adventures to check out too! If you aren't sure of your region's history, keep looking. History lesson complete. Well done.

Maps

If Bigfooting was a game, then maps would be your puzzle board. They are the 'big picture,' so prepare to get acquainted with them. For our purposes, remember we're mad and looking for the unusual, so we need something different. The map we want does not exist, so we will make it ourselves. At a minimum, this is a three-step process. Our primary focus is on the land. Depending on how in-depth you take this hobby, mapmaking can take an indeterminate amount of personal time and energy to get just right. A good map is your most priceless piece of coveted information and will become your most used tool along your Bigfoot journey. So, even entering this field slowly, your map will always be there, waiting patiently. Taking your time is preferable because once you get more familiar with your area, the details will start finding their way there. Since we can't even prove these Squatches are real, let's have fun and start on the basics of mapping your land.

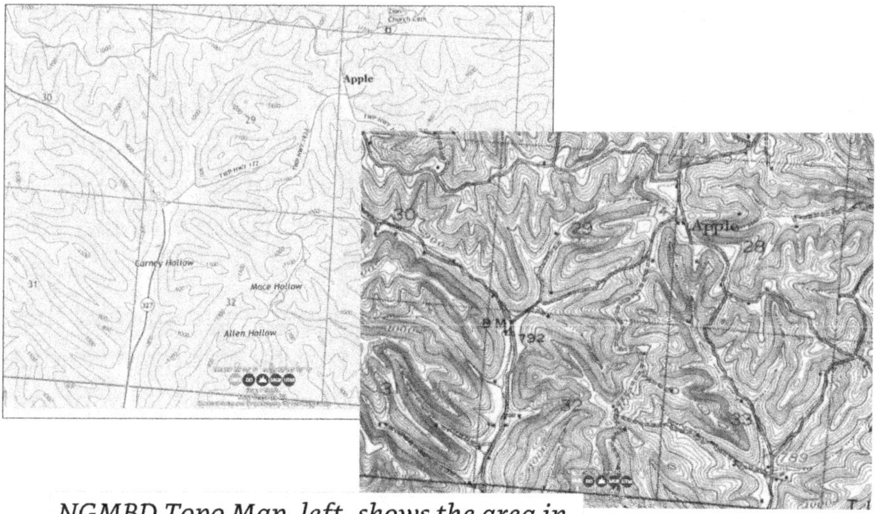

NGMBD Topo Map, left, shows the area in 2019. On right, the area in 1912 when it was farmed and had many residents.

Take time to learn how to read a map, then browse ALL the maps from the topographical to the watershed, forested and reclaimed agriculture, and wetland. Each one will give you different information; some are useful for useless knowledge, while others are important for logistics. The glacier map is always a good one to start with since it formed the literal land we stand on today. What you choose is heavily dependent on your location. For example, if you live in the flatlands or a low valley, you would definitely want the watershed and maybe a wetland and agriculture map for reference. Soils and substrate change based on topography and history, so a topo map would not be a first choice unless you are nearby some hill country. However, it is always a good idea to take a glance at any hidden geographical changes like gorges or cuts.

In cases like John Bryan State Park or Clifton Mills, there is a very deceiving gorge with plenty of historical reports. As opposed, if you live in the hill country, and you found the land was used for mining or clear-cutting, then I would suggest the topo map first, reclaimed second, and the watershed third. Hill country gets a bit more technical regarding habitat because water along ridgetops becomes a strong factor. But these are all just examples. Your job is doing your homework to determine which map will suit you best for your target area.

Now we get a 'scrap map,' a fresh pencil, and stare at this weird piece of paper with lines and maybe colors or shades of grey all over it. It's a beautiful thing, really, a blank canvas just waiting for you to fill in the details. Basically, what we are doing is creating a map key or legend. Remember that everything takes time, so don't get hung up on minutiae right now; stay general. As you're staring at this map, start asking yourself some of those nutty Bigfoot investigator questions.

Specifically, we are looking for two geographical things right now: paths of least resistance and where there are fixed landmarks currently. By this, I mean what areas look easy to get into and what is not changing anytime soon such as waterways, basic topography, and cardinal orientation. Your questions should include the following: What kind of substrate predominates in the area? Where are the pine trees? Where is the water? Where is the swamp that leads to the tributary that flows to the creek? What creek? Are there any points of entry or trailheads nearby? Where are the south slopes? Are there any power or gas line cuts in the area, and in which direction do they go? Does the nearby farm have a natural fence or tree plot? Were there any Indigenous American settlements, mining sites, or old roadways? Does anyone fish in that old pond?

These are just a few of the millions of questions you might think of, and none are wrong. You can invest as much time as you want exploring options, but plotting just a few of these fixed items are the first pieces to put on your map. The answers will come with a little homework, time, and patience. It's good to get into the habit of doing general land mapping before heading out in the field because now we need to get in our vehicles and check them out first-hand!

FIELD TIME! Yep, here is where we finally check out all those interesting places you marked down. Depending on how many there were, this process can take minutes, hours, days, or even years, which is perfectly fine. Bigfooters call this phase 'preliminary scouting.' I call it exploring. If you don't already have one, this would be a good time to make or start a field journal. While in your target region, drive to as many details as possible on your scrap map. Add any offshoot, township, or fire roads you can follow up with later.

Make notes in your journal if you plotted something that looks great first-hand. Add even more landmarks and details because details equal answers. The goal is to ease into your area and get a true perspective. Maps and computers are great, but neither suffices for a first-hand view. When you return home, personalize your map so it will be as useful as possible for future reference. We will add a lot more as we go along, so save these and watch what happens for you.

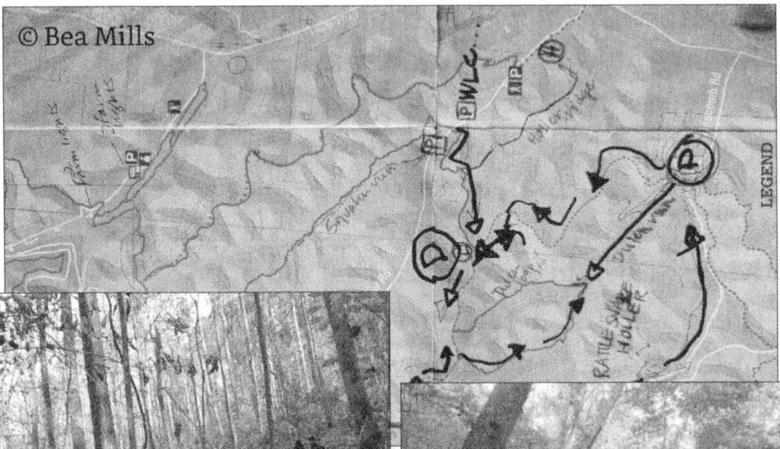

Make a field journal with an area map and take pictures of the area—

Other tools I employ besides a field journal are video and photo notes. It is said that a photo is worth a thousand words. It's true. If you take one picture, take three, and make sure one is as wide an angle as possible, so you capture the surrounding area.

If you are not already doing at least a few of these, know that documentation, by any means possible, is essential for all purposes in life. Documentation keeps the facts in order and can be utilized for review later. Personally, I have several areas I monitor. So, I have all the maps (The more weird and hard-to-find maps are the ones I love the most love, because somebody already took the time, did the homework, and filled in details about the strangest things. I even have maps of two certain counties in Ohio. One shows every old underground mine shaft, and the other, every Indigenous American village and artifact found! Swoon! There is one fun map with crypto and ghost reports on it by an Ohioan; it's called: Map In Black. And yes, I have it too.) I even have a map wall at home that helps me focus on target areas.

For each site, I make up a master map. To do this, I layer all the elements I want to know about the area and combine them into one flat map on the computer. Then I transfer all my notes, scribbles, and whatever else I feel is pertinent onto it. When I'm done dropping pins and measuring everything, I finally add my legend. Now, this mess has effectively become my master and a great tool that I'm always adding and elaborating on. When I need something really specific, like my expedition maps, all I have to do is open my computer and get super technical updating the key.

That's the brilliance of taking the time to make these maps, from running logistics on your expeditions to taking friends out to Squatch or needing your map and compass to get back to your vehicle (which has happened more than once!). It's all a breeze because you've already done your homework. It is a two-fold win that you find your car and know your area. But, because of the time and labor involved, I strongly suggest NEVER sharing your map under any circumstances.

While sharing is usually caring, I recommend avoiding it. Disingenuous Bigfooters can easily steal countless hours, even years, of your hard work. Your maps are for your enjoyment and success in the field. Keep it that way. If you take your time on maps, they will always serve you in ways unimaginable. They are super fun, handy, valuable, and can save you from a really bad day.

PARTS:

Basics

So, we've covered the pieces, on to the parts. By now, you should have a fairly good understanding of your region's land and history and have decided if the target area you chose is plausible and worth your time. To accomplish this next component, we need to determine the relatability of our pieces to their parts with respect to habitat and viability. In determining what actual parts we need, I've narrowed it down to four simple ones, the Basics. Food, Water, Shelter, and Safety. Yes, the very basics, but only on the surface. Each one of these parts has the potential to become a stand-alone college thesis paper, so I strongly advise starting simple. This is precisely where you will often reevaluate when you need to find more information or change directions.

Land is static, whereas our basics can be more questionable in terms of availability. Many of us live where we have seasons throughout the year. To cross reference Bigfoot reports with topographical maps, it's fairly consistent. (Florida is truly unique) For the most part, though, we expect times of abundance and scarcity. Summer drought, winter freeze, spring floods, fall harvest. Other seasons, like tornado season, hurricane season, and Nor'easters. These are basic and mostly predictable patterns. Like moon phases, nature is attuned to patterns and change and has already adapted for survival.

We have the TV weatherman (or some of us are super nerd weather spotters), radio, and National Weather Service Alerts. We have the Red Cross and National Guard.

Unlike nature, we are not prepared for our basics to be unavailable. We talk about our gardens, bumper crops of zucchini, or how bad the tomatoes did, and those fun boil alerts when tap water gets contaminated with mysteries. We have clothes, shoes, coats, food in our bellies, roofs over our heads, and maybe a hug. Those are our basics. However, we're not looking for us, so we can't think like us. Nature does not care. If Sasquatch is part of nature, it would make sense that it is also accustomed to its environment's patterns. Getting in your areas throughout the year is essential. So simple is perfect right now. It won't take long till we level up. With so many new things to check out and learn, don't lose sight of that big, BASIC picture. Even though I expect that, once you get rolling on this, you'll find yourself traveling down all sorts of fun, wild, nerdy nature roads looking for that hairy fella hanging in the woods -yep, I'll be the one fishing by the creek!

Fair warning: This is an ongoing learning process, with field time, reading, and maybe even classes involved. It will become more time-consuming than mapping could hope for, but honestly, it's my favorite part of the entire process. Whatever effort you put into this component will reward you for the rest of your life. Learning only stops when you do. For every slice of new knowledge, there will always be more questions to ask. (And truthfully, you might find that you don't care about Bigfoot at all during this process. But instead, find something else that sparks your interest outdoors. Hobbies are supposed to be fun, so if this helps you find something else amazing to get into and explore, that's fabulous! I could not be a happier gal than to know nature just got a new fan!)

Until then, keep your map and journal open, as we will be filling in even more details on that Bigfoot puzzle board. Let's get full-on Bigfoot Nerdy with my version of Maslow's Hierarchy of Needs.

Water. Our #1 basic. Water is essential for existence, not an option. While finding water for us can be tricky since our digestive tracts are not always able to process the bacteria found in most outdoor bodies of water, adaptations in wild animals can drink with no problem. Water also hosts food. Find the water.

Food. Food is a primary motivator for all living organisms. Since everything needs food, securing it is our #2 basic. Scientists are still learning new things about nature's complex food chain. From Sasquatch to fungus, every plant and animal has a relationship with some form of predator or prey for survival. Find a food source.

Safety. Yes. This is my #3. Perspectives on safety directly trigger our fight or flight mechanisms, behaviors, and compensatory actions and reactions. I have found that safety is highly overlooked and underrated on a needs list, which turned into a KEY basic during an investigation.

Shelter. Nature has a habit of testing limits on everything. Cold versus warm-blooded, vertebrate versus invertebrate. Big animals, little animals. Most animals make or take a form of shelter, while others, like deer, do not. Truthfully, I find shelter to be non-specific in this field. I don't feel Sasquatch are nearly as dependent on shelter as we are, seeking it regularly. Historical reports keep this open for debate on an opportunistic level. To add depth to the list of basic possibilities, I leave Shelter as #4. For mapping purposes, finding shelter areas, like recesses, caves and abandoned structures (besides being creepy), is very good to know.

Map out the terrain—

These basics are the cornerstones that shape the rest of our journey. We've already done the history and mapped out some basics, like the water, terrain, and forested areas. The key is determining if there are enough basics to meet sustainable habitat needs. In essence, now look deeper into your site, and answer more questions. What is edible in your area? Is there enough green to forage plants year-round? Is the water source clean and healthy? Are there any fish, mussels, crawdads, or frogs in there? What kind of soil are you around? Different plants grow in different soil. Are there enough mast-producing trees in your forest to sustain both small game and larger herbivores? Are there any sources of salt or minerals? What was, or still is, in the area that could serve as shelter? Look at your maps for topographical features that could lend to overhangs or recess caves. Have you found any old abandoned homesites or shacks standing in the woods? What kind of terrain do you have that could provide either an easy exit or a pinch point for spooked game? Did you find a map of old mine shafts or see any gated openings during your scouting? (Do Not Enter Mine Shafts - EVER. They can kill you. Just jot the location down for reference like a responsible Bigfooter with common sense, don't be that statistic.)

This is why this component can become very time-consuming. Your list of questions may become endless. The basics are the four cornerstones of knowledge. The details become the building blocks of your foundation. Learning what resources are readily available to support significant calorie needs will help narrow down the possibilities in your area. We're not worrying about trim packages on trucks, or checking barometric pressures yet. This is honestly where I find the most amount of learning opportunities. I circle back here often, looking for things I may have missed and redirecting my focus. Starting out, keep it simple: food, water, safety, and shelter.

Summary

Let's have some fun as I nerd out with an example and sum up this fundamental Principle of HOMEWORK—

I love being asked, "What's with the trees?" I have preached about the importance of trees in Bigfoot since 2014. Trees are, without question, the single most important key to success in Bigfoot and the one thing in this world of madness that makes sense. As a bonafide tree hugger since childhood, this truth took me quite some time to figure out. Once I did, my eyes were opened up to a whole new world of the plausibility of Sasquatch. Now, level up.

Let's assume the average North American Sasquatch is flesh and blood, about 7-8 feet tall, and weighs maybe 400 pounds. It has an omnivorous diet and is related to us. There is (I promise) a calculator that will determine what the recommended caloric needs would be for a person that size. It's called the basal metabolic rate, or BMR. The BMR for a 7ft tall, 400-pound man equals 3,053 calories needed daily to maintain basic body functions.

My regions would be considered young forests, less than 100 years old. Before Ohio's deforestation in the late 1800s, we had glorious native American chestnut trees growing over 100 feet tall, with trunk diameters well over 12 feet wide. Chestnuts were the biggest and most nutritious for the animals and us out of all the trees. Sadly, by the 1950s, they are all but gone from our forests, losing the battle to a simple fungus. Luckily, we have several native nut and fruit -bearing trees and shrubs that have taken back residence in the state. Examples include trees like oak, hickory, and beech and shrubs like the raspberry, blackberry, and blueberry. All these examples produce an edible part that we call a mast.

Let's look at an Oak tree because there are plenty, and you can see them first-hand. Oak trees produce mast which is the acorn. One ounce of acorns is equivalent to 110 calories. Of that, 6.8 grams is the total fat, 12 grams of carbohydrates, and 1.7 grams of protein with plenty of micronutrients. A single oak tree can drop up to ten thousand acorns in one year! This is fantastic, but there's a catch as nature has jokes—there's a whole science behind it. Not all mast trees produce *good* mast every year. Your average Oak only produces a good mast every two to three, sometimes up to four to six years. Besides that, weather and drought make for critical variables in mast production. This is the reason field notes and maps are so important.

Take this one step further, what animals can you think of that eat mast? Answer: Nearly everything competes for wild edibles. Deer. Squirrels, chipmunks, birds, bears, humans, etc. Could a single oak tree support the caloric needs of a 7-foot tall, 400-pound Sasquatch? In short, the answer is yes, it can support the caloric needs, but not by itself. Do you want to get real tree nerdy? Sure, why not. In the United States, there are around 90 species of oak; in Ohio, we have about 14, give or take a few non-natives, producing acorns. It is not unusual to find an entire grove of different oak trees with pin oaks nearby white oaks, and red oaks hanging with the bur oaks. And who knows, maybe a hickory found its way in there by a nutty squirrel. When you find a grove of trees like this, put it on your map because you are nearly guaranteed a consistent fall food source. No matter what year it is, there will be mast with that many oaks in the vicinity. Knowing where the mast trees are in relation to those game trails you've been following is important because of that food chain. Yes, now we're onto something, and you've just landed a decent target area.

What did the chestnut have to do with anything? When Ohio was clear-cut, the chestnut trees had made up a staggering almost 25% of our forests and were now gone. Most of the trees, in general, were gone. Mast-dependent animals had no food to eat. Elk and bison were slaughtered or just moved out. Moving up the food chain - if there is no food for game, there is no game. Why would a Sasquatch want to hang out and starve, too? The timeline gap in Bigfoot reports between early 1900 and 1960 directly responded to this deforestation. It wasn't until nearly all the natural resources were gone that the clear-cutting and strip mining stopped, allowing trees to grow again. A whitetail deer was uncommon in Ohio until the mid-1970s. Once the trees came back, so did many native animals, including our big and quite hairy friend and top-tier predator, Bigfoot.

Ohio now consistently ranks 4th and 5th in the country for Bigfoot reports. While one big tree by itself will feed critters once every few years, the grove of smaller trees, working together, create a more stable environment, providing an abundance of good food annually. Nature and animals know this, and now you do too. So find the grove of trees.

Understanding this symbiotic relationship between trees to land health is paramount. From improving soil nutrients, cleaning water, balancing air pollution, their role in the chain of survival, and just about everything regarding land health, trees are valuable to know. Thankfully, we have groups like the Bartram Land and Wildlife Preservation, the Ohio Division of Natural Resources, volunteers, and many others working hard on land management. Their commitment to forestry health is critical.

Doing your homework will exponentially increase your chances of success. The important part here is getting a better, broader understanding of nature's complex relationships. Figuring out what can readily support significant calorie needs will help further narrow down the possibilities. So now, when you find that great Ohio forest with a history of coal mining, but that big, nearby creek has a pH of 1, you will know that the area is insufficient. Instead of wasting time, you'll keep looking around until you find the tasty water. The basics are usually all you need to get the ball rolling in a very good direction. Keep it simple; make your goals clear, and most importantly, have fun!

I always remind myself of that first day as a high school freshman. This grey-haired, deadpan, slightly hunched, fit man wearing glasses stood leaning over his podium. I'll never forget he drew in a deep, exasperated sigh. Without looking up, he authoritatively declared, "Knowledge for the Sake of Knowledge." Catholic guilt prevented me from laughing. So, I just sat there wondering what I didn't know.

Chapter 3

PRINCIPLE #2: Strategy

A) Time
B) Action
C) Variables

This is where we get into the specifics of field time. Sometimes you'll hear investigators chatting together, bragging about how many hours they've spent 'in the field' this year or years past. They're not bragging. It truly is work, but a different kind of work. It's getting those boots on the ground, cliche but the truth. There is absolutely, positively NO substitute for the field. No amount of armchair hugging, couch sitting, keyboard warrioring, or romantic wishing will take the place of field time. Calling yourself a 'Bigfooter' is easy, but being a 'Bigfooter' is not. Like any hobby, though, which this is, it all depends on you. Bigfoot should be, like any hobby, first and foremost, fun. I love being outdoors, but getting into Bigfoot wasn't exactly planned, and it hasn't been easy. In fact, it all started as a joke, but in the end, the joke was on me.

I grew up all over the place but ended up in the middle of a cornfield with a patch of woods, and a nearby creek that I didn't know existed until I was eight. It was during those days of three TV channels and PBS when the planets aligned. There were only rotary phones and no computers, and it seemed like the Stone Age. But I had tons of books and read everything I could get my hands on. But the nature books and field guides were my favorite because I could take my book outside and look up all the weird plants, rocks, and bugs. I walked the fields and creek year-round, looking for arrowheads and artifacts.

My parents bought me a telescope, and I studied the stars. The woods were my favorite until I learned I was allergic to poison ivy, so I searched and learned about that plant in the book really fast. Since I was the oldest and living in rural Appalachia, growing up was boring. I was a nerd before I knew what a nerd was. When I was 16, I saw a UFO. At 33, I finally told someone about it. I was immediately introduced to MUFON (the Mutual UFO Network) and, closer to home, RUFOS (the nation's oldest group, Roundtown UFO Society, in Circleville, Ohio). Now with computers and the internet, it didn't take long before I began poking around, trying to figure out what I had seen in my teenage days. I started going to the RUFOS meetings and made friends with a few of the folks. I even shared my story and was surprised they didn't laugh. A few years passed, and I was a full-fledged and badged MUFON Field Investigator. It was fun, and I enjoyed it. I loved learning all the fascinating things, from barometric pressures, swamp gas, energy equations, and their respective math to stars, planets, and basic astronomy. Mostly, I enjoyed the investigator part. I loved doing the reports.

Putting all the pieces together to make a case for a potential witness weirded out by the unexplained. I like weird things, I guess? One evening we had a guest speaker who was to tell us about his recent Bigfoot sighting. Skeptical? You bet I was. Bigfoot does not exist except in those books I read as a kid, not in real life. No way. At the meeting, this gentleman told the events of his photoshoot and brought photos to share. My skeptical self could not believe it, not even with a series of resized pictures and a white circle. Yet he pressed on with his story. When he was done, I finally had to ask him, do you believe you saw a Bigfoot? His answer was a wide-eyed "YES!" It was at that point I believed him. Because he believed in himself, having nothing to sell and nothing to prove.

I believed that he believed what he saw. I will not take that from him, not then and not now. So, here's what got interesting—where the event occurred, while he was taking senior photos, was a park I frequented weekly. Oh, I knew this park like the back of my hand. I'd hiked every trail. I had fished, creek-walked, and taken my son there often. I knew the only place to get cell service within a 10-minute drive. I already had taken photos of the area, because it's just a beautiful park. The entire road was part of my unofficial 'clear my head and ponder life' drive. But never had the thought of Bigfoot ever entered my head. It just seemed preposterous. But, as curiosity trumped my skeptical judgment, his presentation was the tipping of Bea Mills down a (now twelve-year) rabbit hole adventure into the depths of Bigfoot, with all of its madness.

His BFRO report was the motivating factor in why I endeavored to become an Ohio Certified Volunteer Naturalist in 2014. The investigator had written a phenomenal report, painting a very accurate picture of the area concerning the event at Lake Ramona. I wanted to be able to speak like that. I wanted to know those details. My old field guides are (still) incredibly useful, especially in this area of no cell service. The internet helped me find answers to nature-related topics, but it wasn't enough; I wanted to know HOW there could be Bigfoot and WHY this area. When I started this, I knew I had a knowledge gap. But I needed to figure out first what I did not know. I foolishly thought I knew a lot until I began the program and got schooled. By taking the eight-week OCVN program, a whole new world opened up to me. I couldn't get enough. I wanted to change careers. I started volunteering everywhere I could, seeking out projects I knew nothing about. The leaders were all fantastic and taught us all sorts of things in the field.

From the importance of pulling invasive stilt grass and garlic mustard to how trillium is propagated and spread. Teaching the relationships of mycology to forest and tree health, and then whole region ecology, and learning the lay of the land. Despite all this, I was still figuring out what I didn't know. Over the last nine years, OCVN has continued to be part of my daily life, and I love every minute of it. Through classes, continuing education, and participating in and leading programs, I've volunteered all over the state, met new people, and been educated on the importance of ecology conservation and preservation. I am the current president of the Hocking Hills Chapter. Because of this extra education, I have been fortunate to fill in many of the questions I had in my field book. I now have the understanding, education, and resources I didn't have. Most states have a volunteer naturalist program available, and many in Ohio have followed my lead to expand their knowledge and appreciation of their areas. By learning, you're looking. Looking firsthand at what your region looks like. Oh, the learning only stops when you do.

Circle back, and let's look again at this park, Clear Creek Metro Park. A lush forest sits on the borderlands of Hocking and Fairfield County, around six thousand acres long.

Clear Creek Metro Park

There are tons of sandstone outcroppings, a clean, beautiful creek draining into the Hocking River, Hemlocks on the low level, and hardwoods on the ridge tops. Migratory birds bring birders in from all over. It is truly a breathtaking area, but Bigfoot? It took years and asking old locals who don't talk, but yes, Bigfoot. WHAT? I continued out on my weekly hikes, but it was funny. I was now hiking and 'looking for Bigfoot.' All the photos I posted were for fun and chuckles because Bigfoot does not exist. But truth be told, I kept a slanty eye open, just in case. And in that spirit of just in case, how did I start this? I started at the beginning, the basics (because I already had a trail map to hike with!) I took pictures, tons of them. I learned about the area's plants and read my field guides. I checked the quality of the nearby creeks; I whispered to the birders while they were recording the migratory songbirds. I looked for game tracks and their trails and wandered along them to see where they went.

I tried finding reports. Being familiar with MUFON's incredible database, I found locating decent information about this hairy guy proved a bit more challenging. Here are a few suggestions for you about your area. Look around for any old reports. This is one of those times when the internet might be useful, although I suggest avoiding it. With some luck, a goodhearted local will offer stories. A few other resources are books; there are several on the market sharing decent collections. If you are lucky enough to find one with local reports, read it. Sometimes, the author might have meetings or information to share that wasn't published. A simple, polite question never hurts. In worst-case scenario, the answer is no more information. Look for any area history of visual (Class A) sightings, road crossings, tracks, and howls. Lastly, it is acceptable to decide that a particular area is a prime location based on your gut instinct. Those are all fine. There is no right or wrong; nature provides us with clues. We just have to know what those are.

Action

So, you've done the 'Time.' Now, what is the 'Action'?

Answer: There are several components to action. And the sky is pretty much the limit. The amount of time you spend in the field creates and navigates your actions. So let your actions become as organic as breathing. This is truly where field time gets fun and the real work begins.

Again, you need to decide what your goals are. And by goals, they might be as simple as going for a walk or as complex as proving Sasquatch exists (which I hope you do!) Most often, I hear, I want to see one! That's great. Then your very first step is to get outside because they rarely welcome themselves into your home. Maybe a barn, not exactly your house. My first sighting was by accident. My second was somewhat intentional but absolutely not expected. The third was terrifying. And the fourth just made me mad. So, I'm good on the sightings here. It would not bother me to ever see one again. I had not planned to ever see a Sasquatch. Call me doubting Thomas or just an absolute skeptic, but up until the second sighting, I did not believe they existed. The first one happened while looking through a thermal for the first time, and I could not believe it. My MUFON investigator brain would not allow for such a radical idea. No way was I watching this mythological thing walk up a ravine. Nope. Not ever. And not even with a true Bigfoot investigator standing beside me, watching the same being. Forget it. (But, I couldn't.)

The second sighting was what ceased all skepticism for me, and it happened in an instant. Of course, by 2014, I was open to the idea of Bigfoot. I liked the folks. We chatted, camped, hiked, explored, and shared campfires, stories, and life. It was fun. I endeavored to put together a private expedition with the help of my boyfriend at the time.

We had been targeting an area he knew well, but I just loved visiting and exploring. Several folks came to join us for the weekend, which I named, 'The search for Uncle Harry'. Yes, it was as complete as I could make it with highlighted maps, an itinerary, meals and a schedule. I had not been on an expedition before, so I just included everything that I would expect from one and had a great time putting it all together.

Besides it being pretty cold for early April, everyone seemed in good spirits. That Saturday night while out on one of those 'Bigfoot hikes', we struck proverbial gold in a moonlit draw in that target area. I watched, what we estimated to be, a 7.5 foot tall Sasquatch walk directly towards me down the slope of the hill. I say walk, because that's exactly what it did, just walk. Arms swinging, gait easy and forward. I locked eyes with this hair covered football player looking species of unknown origin about 20 yards away, and my mind just went blank. I might be a quiet natured person, but I am never speechless. For this event, and to this day, I still have no words to describe it. I had no feelings, and no words. I simply lowered my gaze to my boots and told myself, I can't see this.

In the few seconds it took me to do just that, I looked up again expecting to see this beast right in front of me, but instead it was gone and there sounded what I can only imagine an elephant crashing through brush and trees directly into water. I still just stood there, completely dumbfounded at the entire event. It didn't last more than a few minutes, but I cannot unsee what I saw. It was in those brief moments, I passed the point of no return down the rabbit hole. Now I had to know more, and decided to start putting in the dedicated work. The joke was over, Bigfoot does exist - at least to me.

It was from that life-altering event I actively started minding my 4 Principles. We will call the previous years all 'practice.' And I was never so grateful for that practice, which became invaluable leading to the biggest investigation thus far of my life. So to review: We gather our pieces and parts- history. Print, order, or pick up a topo map (Getting your field journal or notebook would be a wise idea now. And honestly, if pen and paper aren't your cup of tea, find a phone app that doesn't require cell service.) Start learning everything possible about that particular area and make sure to include any suspect or known (Bigfoot) activity. Get IN THE FIELD and start marking off the areas and trails you take. Take notes, photos, and videos for reference. Chart landmarks, powerlines, pull-offs, dumpsters, and open structures like barns and sheds. The key point here is anything you can think of to help you remember the details. I use a form of shorthand in the field, which is why I don't bother sharing notes. To your average left brain, it's all nonsense. For me, it's perfect. Take notes! Learn your history, know your basics, get to the woods, and start Squatching!

STRATEGY

Three Rules of Squatching:

Rule 1) You are always going up a hill

Rule 2) Anything that can go wrong will go wrong.

Rule 3) It's on Squatch's time.

- Mark Maisel

How does one even pretend to implement a strategy? Great question. Glad you asked. Strategy, in this respect, is taking all of the knowledge you've gathered and formulating your ideas on 'best odds.' We practice strategy in all sorts of different ways in everyday life, from simple decisions to careers, future plans, what roads you take home, you name it. Sometimes, they work out, and sometimes they bust. The important thing is you did your best with what information you had. This is no different, except implementing strategy in Bigfoot is like herding squirrels, so don't get discouraged. It is truly hard work and a bit of luck. While most of these are akin to field time with multiple people involved, strategy by yourself is easier. It affords you the time needed to take decent notes and photos without distraction. Depending on your personal goals of attending or leading an expedition or staying solo in your area, these are all good suggestions.

Variables:

As we get technical in the field, let me touch on two variables that influence strategy almost disproportionately: weather and surrounding noises. Through my notes and experiences during investigations, when formulating strategy or collecting history, these two factors are key variables to know. Do not forget these in your investigations.

Weather & Surrounding Noises —What on earth do weather and noises have to do with Bigfoot?

Almost everything. Aren't these basic? Still, feeling basic? Remember, we're mad and now strategically looking for a mythological beast roaming the land, right? So let's act like it. Let's throw a few more variables into the strategy narrative to explore.

I hesitate to use the word pattern because there is no appreciative 'Bigfoot' pattern to lay out. Nor will I say that one does not exist; I said appreciative. However, there is most certainly a hypothetical pattern that I willingly employ in the field based on my own experiences. If a good investigator asks, "What was the weather immediately before, during, and after a said occurrence," they can divulge all sorts of information. I've found that several events occur just before a large weather change. Since 2014, I have learned a lot. During a long-term investigation in Fairfield County, Ohio, the Sasquatch vocalizers I studied were like clockwork before big pressure changes like heavy storms and snow. In the dry seasons, I noted a decreased amount of recorded activity. During the wet seasons, they were just the opposite, heavily increased. In 2014, I co-investigated a thirty-two snow print track line in Scioto County that a witness' dog found at a very rural rest stop just minutes before an ice storm blew in that paralyzed several counties (BFRO report #44171). In 2019, one full footprint and one half footprint were found in April between a period of stormy weather. The thirteen-footprint trackline I investigated occurred during an unseasonably wet July. It became apparent that hard winters made for open and obvious class-A sightings when several private reports came in over a short span of time, all within a few miles of each other. My best theory was desperate foraging, as the timing was during a long, frigid cold snap. I quickly discovered (by accident) that a nuisance kill pile was in a nearby field.

During times of actual precipitation, all activity would cease. I would also suggest that their behaviors resemble crepuscular wildlife (dawn and dusk), with occasional day and nighttime events. Wicked hot days led to activity later into the night. Cool days had earlier activity. The observation list is long, and these are only a few to mention. I feel, however, they are worth noting. *Weather*. Ohio's weather is as unpredictable as Sasquatch, but tools are available to help 'forecast' such things. Utilizing the NOAA website is a great resource for historical weather in the rare case that you just forgot to jot it down.

My second variable would be noises. Noises are just that, obtrusive, obnoxious, and loud. Try it right now, wherever you're located, and just let yourself be still. Listen to your space. Tell the commentators in your head to shush for just 30 seconds. What sounds do you hear? What sounds do you know, and lastly, what sounds do you NOT know? This is a good practice I've borrowed from birders, and suggest giving it a try when you're out in the field. If you aren't already familiar, check later for any nearby airports, active EMS, fire, police stations, working farms, wineries with entertainment, or even just loud neighbors around. Coyotes roam everywhere. Packs are sometimes heard barking and howling before or during Bigfoot activity, almost like those noisy neighbors. Don't discount the turbo screaming on a diesel engine or the crotch rocket flying at Mach 7 down the country road. Plenty of other sounds, like knocks and whoops, have been recorded and reported after sudden, loud noises. For example, I heard fantastic knocks immediately after slamming my car door shut after arriving at one of my locations. (Always start your recorder before getting out. You never know!) It was followed by two beautiful handprints left on my car when I returned. I realized later that I'd left half of a Subway sub in my backseat!

I can honestly summarize, from my observations, that these Bigfoot are somewhat delicate flowers and quite perceptive regarding their hearing. It's almost like they have quite a distaste for noisy interruptions to their sounds of nature. Hence, when out in the woods, I've found it perfectly okay to be yourself, not silent, not noisy, but aware and always listening. I am forever stopping to look, photograph, and listen. In those pauses, you often hear things that 'don't fit.' And there are all sorts of strange sounds in the field, like grass after rain, that make a funny, popping noise to trees bursting in the wintertime from quick drops in the temperature. Listen for the whistles, grunts, growls, screams, and hissing sounds traditionally identified with some extra work and deductive reasoning. On more than one occasion, I've had to reach out for expert opinions to help with identification, from fox screams to nearby gas wells to migratory birds and everything in the middle. Fair warning, it's best if you're able to record that unidentified noise because trying to imitate a call or sound to someone always turns into fits of sad laughter, trust me. It's okay not to know; I admit I don't know everything. Ignorance is not an excuse for many things, especially when it comes to noise. Find your tribe of trusted consultants and those with the knowledge to help teach you. My tribe comprises bird, canine, cat, bear, industrial experts, a game officer, park rangers, naturalists, two hunters, and occasionally, the local neighbor. If they can't help me find the answer, they are always willing to help me find the right person. Suppose you are surefire positive that you've got a Bigfoot recorded because you've exhausted all other possibilities. In that case, there are folks who will give it a quality, thorough listen. Search online for Sasquatch BioAcoustics.

I do not knock on trees or attempt calls unless I am bored and about to leave the field. I find them useless. There are many theories on knocks and calls, and while I have heard them and believe them to have been purposeful and not man-made. My question is the same as yours, why? I can only theorize that they are forms of communication, but what the communication is, I do not know. Are they cool to hear? Absolutely, and definitely worth jotting down a general area on your map. But do you need to knock back? Well, that's up to you, and if you have company. I suspect that by the time you hear a knock, you've already been found out, and the game's about over anyway. However, if you have company or teams, this would be a good time to actively listen, move folks around in that direction and knock back.

One thing I can definitely say is this: different regions seem to favor different Squatch behaviors. For example, you might hear an investigator say to you, "Listen for the knocks; they're knockers there." Or, that area has whistlers, rock clackers, howlers, whatever. While I had heard this myself for years, it wasn't easy to wrap my head around it when attempting to determine a possible range around my investigation site. From what evidence I had collected over the years, I surmised a series of "expectations". In that, the behaviors I was already familiar with in the field would be, at the very least, similar. Now, any Squatchy activity is good activity. But honestly, I was disappointed when all I was catching in my targeted range were incredible knocks. Completely at a loss, I started asking investigators all over the country if they had noticed different behaviors in their different areas. Their conclusions were *yes*, but no one had an answer as to why- it just was what it was. Although appreciated, it was not a good enough answer for me. I am still determined that there is a reason for the madness, and perhaps I just wasn't looking at my data the right way.

So after staring at my maps and wondering where my 'guesstimates' went wrong, I finally came up with a theory. Where's your map? It's the terrain. I decided the localities terrain is the determining factor in regional behaviors. How? Try this experiment for yourself.

Grab a recorder, or if you're lucky, grab a few bored friends and head outside with a notebook (and rangefinder if you're feeling fancy). Find two areas of different terrain with different landmarks. Make a note of the season and weather as well. To really explore this experiment, it should be done in both full foliage and bare woods.

Then, yell (or call blast) in a valley and listen for how far that sound goes. Now, whack a poor hardwood tree like a pro ball player and listen again. Then, repeat in wider valleys or even flat land.

You will notice a distinct difference in how the sound travels. Time, weather, wind, humidity, foliage, and water also directly influence the distance carrying capacity, but terrain wins and can be quite deceiving in hill country.

I use this experiment early in my newer areas where I don't have much background information to work with. Think to yourself, if you were in distress with no phone and needed help, what sounds would get someone else attention best? While I will not say, by any means, that this is a surefire way to find or call in Squatch, this is my best theory concerning regional behaviorisms. To me, it just makes sense. It is worth taking some time to determine what sounds would be most advantageous in your area. And keep adding noises in your notes.

More strategy. We know our history of the land, gathered our pieces and parts, explored the basics, considered key variables, got in the field, and narrowed our region down to a plausible target area. You have done FANTASTIC! I bet your map is looking good by now.

Here are the three keys to remember in strategy:

1) Prepare to give up the high ground

2) Find and follow the water

3) Your attitude is everything

WHY on earth would I give up the high ground? Because, tiny human, when speaking of Top Tier predators in relation to a Sasquatch, you are not at the top of that list. Do I classify a Sasquatch as a predator? Not particularly to us as humans; however, from my own investigations, hearing, and reading many reports, I would not go out of my way to upset one. I've spoken to witnesses with stories from wrapping a full-size deer around a tree with one hand to breaking spines in half, so yes, I would consider a Sasquatch an apex. In any fashion of hunting, even from just watching animals, they customarily find an advantageous spot with a good line of sight and wait. But I don't feel that we are in any position to 'hunt' a Sasquatch. If anything, they are the ones watching us.

I do enjoy the occasional reports where the witness seems to have the element of surprise; however, in my experience, they predominantly know when a loud, strange-smelling human is in their backyard. That said, I traditionally practice a passive role in my target areas, adopting the character of a wandering solo hiker. This is where your topo map comes in handy. If I have reason to believe that a Squatch is in the area, depending on safety and common sense, I give up the high ground. I highly recommend taking the half-slope trails when available and keep looking up. Playing this role in the field has proven to give more opportunities for events to occur. From knocks to visuals, moving slowly and giving Squatch the advantage has paid off in dividends.

WHY follow the water? Re-read the basics—three reasons - Everything needs water. Most aquatic life is edible. And lastly, waterways are usually paths of least resistance, depending on your terrain. Don't discount swamps. Swampy areas are teeming with life, food, and water and have hosted a fair amount of reports. The majority of tracks I have ever found were along or in the vicinity of waterways. Water is a survival need, and animals travel. Depending on your geographic location, water sources are composed of different substrates. Some creeks and rivers may be very rocky, while others are pebble-like and sandy. Some ponds are surrounded by hard, packed earth, while others may be swampland. Depending on your goals, if you are out looking for tracks, checking the sand bars along creeks and rivers is ideal. Check the areas surrounding the water source if the weather has been wet or your area's soils are consistently damp.

While out, look for what I call supporting clues. These would be additional animal tracks, like raccoon, beaver, and deer. If you happen upon a nice game trail leading to a water source, go there and prepare to do some work. More work? Yep. Based on my experience, I would suggest that the average step for a Sasquatch is around 65 inches (5 feet, 5 inches), with the average stride being 90 to 120 inches (a maximum of 10 feet). The work comes in now. Let's assume that the heavily traveled game trail leading to the water has plenty of deer tracks. You feel that it's prime and worth devoting some time. Then take the time to check the periphery. If you are unsure of your own pace, making a line to measure distance is easy, and weighs less than a rangefinder. Start by cutting a piece of cord about 15 feet long. Then mark off lengths of 5 and 10 feet. For accuracy, do not start at the beginning of the cord (because you might need to tie it off on a branch).

Mapped trail—

From your game trail or any point of origin, measure out 10 feet perpendicular to the trail. This becomes your hot zone, and now you know where to look for potential tracks paralleling this game trail. Observation skills are necessary here, and the time of year can make this task easy for future reference. Pan around for marker trees to keep your borders tidy, and look to determine if one side of the path has more or fewer obstructions. You can get a good look at the land after the fall season and before the mid-April leaf out. These game trails are good to insert onto your trail map for reference. Once you become familiar with distances and your area, these will be the area's to poke around and hopefully strike gold. I will not guarantee anything, but I will tell you that this has been a consistent method for area track finds.

What about my attitude? This is my favorite. Your attitude is the key to everything in life. Honestly, the odds of coming upon a Bigfoot are slim to none. No matter how much you plan, how many hours you spend in the field.

No matter what the situation. Bigfoot doesn't show up because you want it to. That's not to say that it doesn't or will never happen; it's just that the odds are incredible. Disappointment is real in this field, but it doesn't have to be. If you go storming into the field, all bent and bound to find that Sasquatch, I bet even the birds will avoid you. If you take the enjoyable, passive approach, your odds increase substantially. You appear non-threatening or hostile in this way. Since this process is multi-faceted, there are plenty of wins in the field, besides getting rocks thrown at you or a high-five from Mr. Howdy.

Keeping your expectations to nothing more than calling it a 'nature walk' or 'scouting' will keep your motivation in this field of madness way up, preventing burnout and grumpiness. Perseverance has paid off for many investigators with whom I've talked this through. Several were in prime locations but getting no action. It took nearly a year for one to get any Squatchy action, but it paid off in the end! Don't give up, and learn to enjoy the little things in nature, because Rule #3 is right every time.

Chapter 4

PRINCIPLE #3: Execution

A) Case Study

B) 2019 Track Investigations, Fairfield County, Ohio

C) Methodology Review

Execution is exactly that, carrying out a plan. In the first two principles, we built the framework to make that plan; now, we have to implement it. Being prepared for life requires an extraordinary amount of work. Whether it is a new career, academic studies, starting a family, moving, or even buying that truck, having solid background knowledge to make good decisions and a working hypothesis often makes these challenges easier. Consider this portion the big final of your school or class before the thesis is due.

Everything we have put together comes into play now with any activity that warrants further investigation. For example, I had no idea what would happen to my life when I first heard the phone call recording from a witness who had an extraordinary claim of very aggressive bigfoot activity on his property. Although I'm typically a pacifist when situations become tense, my skeptical ER Nurse brain usually sends out red flags when stories sound outrageous. I was smack in the middle of a huge life transition with my son graduating high school and getting my house ready to sell. What I expected to be nothing more than a local site investigation and a follow-up report has changed my life forever. Giving due diligence to properly investigate every claim and every piece of evidence while trying my hardest to make any sense of this active property has consumed years of my life, complete with blood, sweat, and tears.

I pushed the envelope on every principle and had to add multiple components to everyone. The more I learned about the area, the more I needed to know. As the investigation continued to heat up, I found I ultimately had one goal, and it was the correct goal: To gather as much information as possible because the odds of something like this ever happening again were about zero percent. I felt an incredible amount of pressure, as this case was unprecedented.

Remember those four groups from earlier? I knew I would be answering every one of them, so I looked to each of them, hoping to glean any insight from their rhetoric. I kept this investigation as low profile as possible to preserve its integrity. Given that the location was so close by, I was even given a moniker, Ghost, to keep my name out of it (Which I don't recommend ever doing, for the record). Every opportunity I had to document, I did. Every piece of information, no matter how ridiculous it seemed, was noted on the off chance that that one detail was an answer I would need later.

After years of wandering around different regions around the state, I felt like I had finally been given one shot at cracking this bigfoot case, nearly in my backyard. I wanted to do the best I could most systematically so that the next generation could look back for study, and hopefully find the clues or parallels I missed, and do better than myself- that was the end goal. After that first site investigation, I knew this was going to be a challenge, and within a few weeks, I knew I had to decide—either commit to this game of Bigfoot or hang it up altogether. Morbid curiosity won, and like those old 'choose your adventure books' from ages ago, each path I explored during this investigation led to different answers with more questions.

What I truly learned from this unexpected twist in my life was; Bigfooting is pure madness disguised in camouflage and conferences. It should not be a competition. It's a hobby, and we are all explorers. So, allow me to take you on just a few of the mad adventures that followed that witness' first phone call.

A) CASE STUDY:

Now, let's review a real investigation examining how I utilized the four principles—*Homework, Strategy and Execution, and Resolution.*

Included are photos, investigators' notes, and the report, followed by my thoughts and a summary. I also included anecdotal information that occurred during the events. While this does not begin to cover the entirety of the multi-year investigation, I do discuss, in detail, some significant events that occurred during the year 2019.

History: In 2016, as an investigator for the Bigfoot Field Researchers Organization, I was appointed lead on a very active piece of private property. This was my first experience with what turned out to be a supernova of information. The report was about a family terrorized by an alleged Bigfoot outside their home. Howls, screams, and objects were being thrown at their house. Chickens went missing, and their pet dog had strange injuries, both internal and external, and eventually passed away. Neighbors' calves were found dehided and stuffed up in trees, with their hut destroyed and found over the cattle fence. The family's mother hid in a locked bedroom with her children while the father would be outside shooting into the darkness. Despite all their best efforts, the father could find no resolutions to deter whatever he was dealing with. Over two months during this series of events, not only was the entire household upset, but so was the surrounding area. It was then that the property owner reached out for help.

Late June 2016 was the initial site investigation (which, incidentally, had immediately followed a harrowing night for the family). When I arrived, a family met me, asking if I was from the TV show, Finding Bigfoot. Clearly, I disappointed everyone as they just stood there, stared at me, and eventually walked away. A few minutes later, the obvious property owner came briskly walking from around the back of the home's garage. He was dressed in full fatigues, including a bunratty. Pistols strapped to both hips, a rifle slung across his chest, and three large knives strategically secured. Obviously exhausted and red-eyed, this was indeed my witness.

I didn't get a chance to bother with formal introductions; he looked at my BFRO polo shirt and immediately started recounting the events from the previous evening. We began walking as he gave a step-by-step tour, talking so fast I just stuffed my notebook in my back pocket. He pointed out the tall farm light on a small slope connecting the garage to the top barn where a chicken coop was in active use, illuminating the entire area, including the tree line just behind the house. Then another spotlight on the garage that brightened even more of the vicinity. We walked up to the barn, then the chicken coop, and proceeded behind the barn, where he stopped to show a second pen of pullets (baby chickens).

The chicken pen—

He explained that this was where he initially saw what he thought was a chicken thief leaning over and into a 6-foot-tall fence. As this man continued talking, we walked to where he ran after the 'thief' into the woods and kneeled to take a shot, realizing that the culprit was no human man running on two legs. He pointed out the logs thrown at him, never getting a round fired. Then, he pointed down the slope to show the big find of this family's terrifying night- A single footprint laid smack in the middle of a small sandy stream of a narrow draw. Upon closer inspection, the print was identifiable with proportions appropriate for a Sasquatch, although not castable.

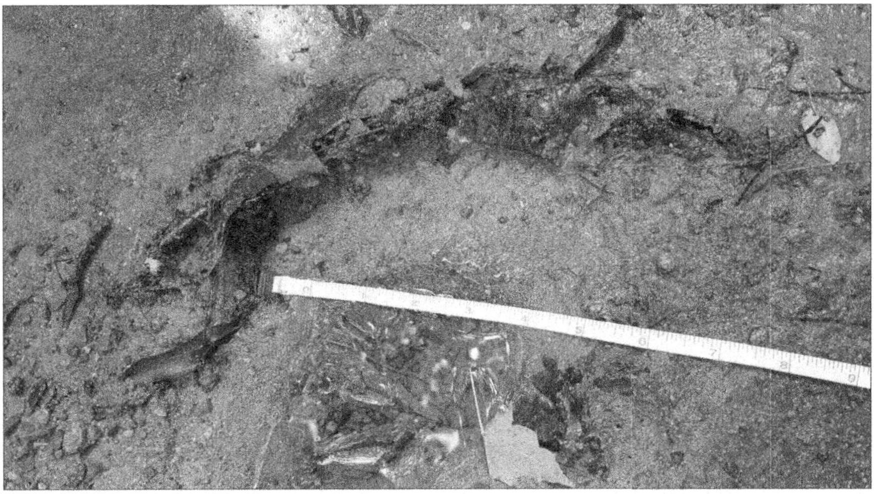

Returning to the driveway, the witness pulled out his phone and stopped for a breath. Despite all the going-ons, he had to wherewithal to capture two recent video recordings of alleged howls. One was meh, but the other was impressive as he had opened the window and stuck his phone to the screen. With his permission, I recorded it onto my phone, but it was useless, the quality was below sub-par, and the file was too large to text, so I was at an impasse. I took a minute in the blistering hot, humid sun and finally scribbled a few shorthand notes. Meanwhile, the witness, peering at me intently, asked, "What are you gonna do about it?" I just stared back, having absolutely no idea, and answered, "Try to figure this thing out."

I got a copy of his video evidence and sent it off to be analyzed. I had never heard a Bigfoot howl besides on my computer, and I was skeptical of this investigation. While the print was classic, the report was quite remarkable. So much activity, so much aggression. I tried to rationalize the situation but kept coming up short. If this man believes his family is being terrorized by Sasquatch, then why? I didn't have enough information, and in no book could I find a report this unique, this active, and with such a duration.

Independently, each party made sense, but I could not find what connected it. So I had to go all the way back to the basics and repeat and repeat and repeat until I could formulate a good working hypothesis for this investigation. The audio analysis reply came back a few days later and was agreeable, noting that the recorded vocalization met the parameters of what we call a North American Sasquatch. It was at that moment, from this feedback, that the entire investigation changed. Boots on. First, I set up a list of goals and zoned in on two big ones. The first was to collect and capture all possible evidence while the site was active. And second, figure out the most logical reason behind the aggressive behavior to alert the family for possible solutions. Now that a non-interested third party had vetted the vocalization, giving credence to this story, I started a recorder project within the week.

Investigations are dynamic and require thinking out of the box. Three things must happen to maintain an area's integrity and any homeowner's right to privacy. Keep your mouth shut, and eyes open, and never give out private locations. We have an obligation to protect our witnesses' privacy no matter how 'hot' or 'active' an area is. Trespassing is illegal. Through disingenuous investigators and neighbors, people started harassing the witness family daily, demanding access to their property. And some, armed with weapons, just trespassed and returned to their vehicles, meeting the county sheriff's officer. Others hung recorders nearby and never said a word until they lost their unit. The list of truancies is long. The list of madness is even longer. Never give out your locations. Never share your maps.

Getting started on recording, I first went to the local electronics store and bought a $30 Sony audio recorder that lasted about 7 hours per night and two triple-A batteries.

I didn't know where to put it. Since I had nothing rigged to be waterproof, I kept it in their three-season room. The mic quality was only good for short-range and not practical for what I wanted to capture. It was a nice idea and a small investment to start. I would recommend this unit for hikes and overnights but not for heavy, long-term fieldwork. There was no acceptable audio recovered using this device.

The vocalization reports were constant. More chickens were missing, and objects were still being thrown at the house. Nearly every night, I was getting a phone call or a text asking me what I was doing. It was just chaos, and I had nothing to show, nothing to help, no answers, and no idea what I would do. If you believe in divine intervention, I finally got on the internet. I began following a group thread where other Bigfoot folks were doing audio recording work and getting the results I wanted, not just a bunch of echoing noises. Then, the Tascam came up. I also upgraded to external microphones to the Tascam Dr-05 from the big city

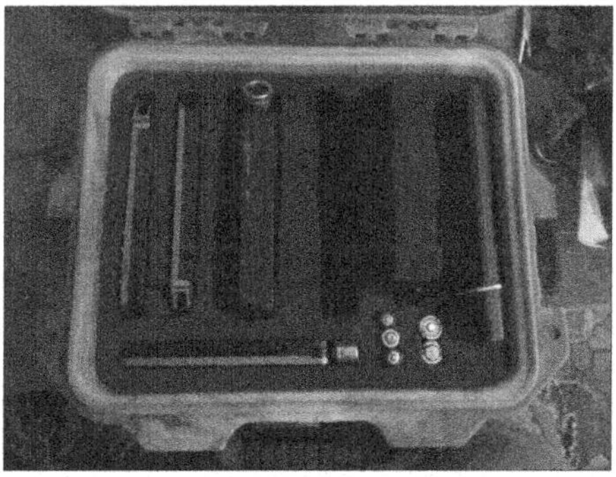

music store. The total cost was $150, and one ponytail holder. Now the unit could be placed outside in a better location, being more weather resistant. It ran on two double A batteries and recorded about 20 hours, depending on the weather. I did field trials of the Tascam Dr -03 and 07, which were rocky at best. Even though both were programmable, running on batteries alone, the recording time was only a few hours. Using an external power bank proved useless to me with faulty connectors, and the battery bank automatically turned itself off. I stuck with the Tascam Dr-05 for nearly a year and still use it for overnighters and at home. This meant I was back to changing batteries every single day.

And every day, the property owner asked me if I had my gun on me, and every day I said, "No. Why?" And every day, he told me, One of these times, you're not coming out of the woods alive, and every day I did just that. I made this time as productive as possible to amass knowledge and record several field notes from the electronics to the property's animals, birds, and insects. I learned every felled tree, chipmunk hole, quicksand spot, and recess cave. I knew where the trail cameras were in relation to the deer paths. I found out where the mast and fruit trees grew. I memorized every bend in the three streams, the water quality, and their respective sources. And I mapped them all out.

By the end of summer moving into fall, the investigation had become frustrating. Being on that property every evening, changing batteries, and walking the streams, woods, and fields was beautiful. I could now hear these incredible vocals from the recorder, but nothing was happening in person. My field notes were nature related, and I was getting burnt out on this investigation already. Then one evening, another investigator and I were standing outside along the nearby tree line with the witness when it happened. If you have never heard a Sasquatch vocalization, there is no way to explain it truly. I finally heard one- real-time, in real life. I felt every emotion I think a person could feel all at once. My mouth fell open, my eyes went wide, and I literally broke out in tears. This powerful bellow came roaring out of the woods and bored right through me. It was indescribable, almost reverent. I felt every fiber in my being tingle, and my hair stood on end. Finally, I got to hear what the family and that damn recorder were hearing all the time - a perfect Ohio Howl! I never wanted it to end. I've seen Motorhead in concert, but this was just WOW, and my motivation returned. I hope someday you get to hear a Squatch howl too. Back to work.

Finally, the answer to a true long-duration recorder became available, the Olympus 620 DM. I ran it for 9 hours a night safely for 28 days at a time, full throttle. These recorders did their job, and the success was immediate. Life

became much better. Hundreds to thousands of hours of audio quickly racked up, becoming more than overwhelming. I had spare time now and I spent it manually listening to each and every minute of data while driving in my car, cleaning the house, and taking breaks at work. Anything of weird significance was recorded onto my phone and sent out. Yes, it was insanity, but it worked, and it was all I knew how to do. Insert technology. It was my son who came to the rescue and taught me the basics of an audio program he used for his album recording, Reaper. (Audacity is a free alternative to Reaper. I had a fantastic tutorial from a colleague and co-author of 'Squatchin' 101'. Additionally, I highly recommend checking out Sasquatch BioAcoustics on the internet. It just wasn't a good fit for me, though)

The few folks I shared the audio with agreed to their validity, yet the obvious was becoming more obvious. Where are they? Yes, Bigfoot is in the area, but why haven't I found any tracks, hair, or scat? This was even more frustrating for me. It wasn't as though I wasn't looking. Three years had passed now, and all I had to show was one tiny track that looked like a mashed-up ball of plaster cast from a bone-dry field that was older than dirt.

All the surrounding properties are private, and most of those folks kept to themselves. So, back to the basics. The grounds on the property were not great for tracks. Thick leaf litter in dense woods that immediately changed into a horrifying tall reed swamp at the tree line does not make the best substrate to 'find tracks.' The creeks were, for the most part, narrow and pitched. When the rains came, they immediately flooded.

Eventually, when asked about additional evidence, I just shook my head and shrugged my shoulders. In short, I found nothing worth showing except photos of possible tracks. Who wants just photos? People want plaster, not possibilities. I get it. I wanted to find something worth casting, not depressions in the reeds or smashed-down leaf litter. My self-pep-talks after these conversations went something like this, "It's Bigfoot, and we can't even prove it exists.

How common are finding tracks anyways, especially in Ohio? It's not, or the entire country would have tracks to share." So I resigned just to be happy with the audio and continue working on my data library, reminding myself that any mere attempt to explain this investigation to anyone not mad like me comes off as ridiculous anyways. Ask my co-workers!

There was one occasion of another enthusiast who lived (as the crow flies) approximately three miles away whom I kept in contact with often as he kept an intermittent recorder going. He would send me audio occasionally, which matched my vocalizers. However one instance, and it never happened again, we had the same vocalizer recorded simultaneously. I was blown away. A three-mile difference on the same recording was a real eye-opener. It also increased the range exponentially and, given the lay of the land, made my search for evidence harder. I knew it would be near impossible to find hard evidence; I know better than to trespass on others' property and I don't lie to get access.

But it did lend many clues, and I had to think even further out of the box. So, I changed vectors a bit and started learning everything I could about how sound travels. I asked unsuspecting people the most ridiculous of questions, and while most of them laughed at me, it was a start. I had a new goal now: to try and figure out a better direction. Many things happened during this long-term investigation, but most are omitted because only time and resolve can properly write it out. But this next story is exactly why I added safety to my list of basics, and I believe it answered one of my initial goal questions explaining early onset aggressiveness.

Now, clues show themselves in the strangest ways, and to this day, August 22nd, 2016, still remains a mystery. It was on this night that the dynamic of the entire investigation changed, specifically to the pattern of aggressive behavior. Recently, I moved my Tascam Dr-05 to a new location. It was now down the hill in the woods, hanging up in a tree about fifteen feet alongside their four-wheeler trail. From the three-season room, maybe 100 yards distance. Having just maintained the recorder with new batteries and an SD card, it was nearly dark when I left the property in my noisy engine-knocking Subaru.

I hadn't made it three miles down the road before my phone rang. The property's father was yelling at me to turn around and come back. With my arm outstretched, holding the phone away from my ear, my skin began crawling, hearing his words because they all had just heard the loudest, longest, most impressive howl since this all began.

And it was immediately behind their house!

I didn't tell the family that, while checking my rig earlier, I had gotten this overall feeling of tension. I felt I was being watched, and for lack of a better term, the heebie-jeebies, and just wanted to leave. I don't announce feelings like that, being completely subjective, but the phone call didn't exactly surprise me. Already uncomfortable, returning was *not* on my priority list, but I did what I signed up to do as an investigator; I turned my vehicle around. I met the family in the driveway and asked again what had happened. Same story. We settled in the four-season room for a few reasons. One, it was 100 yards from my recorder. Two, the room was made up of all windows and had steps leading down onto another open floor, in a split-level style, only it was all open with high ceilings, so the acoustics rolling up from the swamp and forest valley immediately below, echoed quite well into the room.

The vocals started up again while standing there, only they were further away, more distant. I went to my car and returned with the backup sony recorder, turned it on, and placed it by the open window. I pointed the mics so that the family were oriented towards the vocalizations. With much hesitancy and reticence, I told them I needed to get down to my recorder in the woods. They laughed at me and said, "Have fun." In my mind, I was confused and somewhat in disbelief of their story, requesting me to turn around. I wanted to hear what they claimed, and I wanted to hear it immediately. It was not a question of was it Bigfoot, but WHY? I wanted to hear if there were any clues in that vocalization, being that it was not even 30 minutes old. The father handed me a pistol and a fancy flashlight as I stood. Outside, standing at the top of the trail, I looked down at that dark forest. Although the tall farm light was positioned just behind me, the tree canopy was so thick that I was looking at total blackness.

It was dizzying. There wasn't a sound heard in those moments. I heard no cars, air conditioning, insects, or planes—nothing. I turned on the fancy flashlight, which still didn't make me feel better, growled at myself, and took off sprinting down the hill. No, it wasn't pretty, but I got down there. I dropped the flashlight, untied the cable, grabbed my entire rig, and ran back up the hill. Yep, I lived. And I didn't care if a herd of Bigfoot laughed at me. A little over an hour had passed since I hung this device which was working properly. However, when I opened it up, it was off. The batteries were completely dead.

Thinking back in my head to my own 2013 and 2014 experiences when all the electronics died, I looked at my dead recorder lying in its case and thought, well, this is interesting. The Tascam recorded about 25 minutes of audio before pooping out. In that data, just 14 or so minutes after I had hung it, a complete four-minute vocalization, only pausing to breathe, was recorded, spiking so hard it muted the mics a few times. Okay, I had to eat crow on that one. It was impressive, but now my good recorder was out of commission, and the only one I had running was the Sony inside the house. Not good. I replaced the batteries and changed the card on the Tascam, on the off chance something was corrupted, to save the audio I did have and moved it outside to their child's playset. While I was standing out there, it wasn't until a few minutes later that all hell broke loose in their valley.

Every dog in the entire region started barking. One vocalizer began howling, then another, and then a third. The last one sounded different, though. It was this horrible howl/wailing. I turned, and we all stared at each other. The wind had picked up, now a slight breeze, and the vocals were traveling on it. As the wind was coming from the northwest, that was the direction I looked. It was incessant, this last vocalizer, constant and loud. The other two were just as loud but were the (now usual to me) howls. All three seemed to be in different locations; nothing made sense except that something big was happening. In those few seconds of pondering, I looked around the countryside. It looked like something out of a movie as I watched the surrounding houses start turning on their outside lights. I heard neighbors beginning to yell at their dogs, then at each other, asking, "What is that noise?"

As the wailing and howling continued, I felt panic growing inside me, my hair standing up again, and then it hit me. Being a woman is weird, and we women experience all sorts of strange things. Coupled with my nursing career and years working in the emergency room, there wasn't much weirdness I hadn't experienced firsthand. I turned again to the family and announced, "I think one of your Squatches is giving birth." The father bolted into the house and grabbed his keys, ran back outside, and yelled to me, "Let's go!" He fired up his loud big, green diesel truck and rolled the windows down while tearing out the driveway. When we got to the road, we turned left to follow the direction of the breeze. We stopped for just an instant a few hundred yards down the road.

The spot was advantageous—clearing the tree line, opening the line of sight, and being on higher ground. Even over the decibels of the diesel engine sitting idle, the wailing could still be heard. I had not grabbed a recorder, so I figured the next best thing was to start recording on my phone.

We drove up a lane on the opposing ridge and passed the row of houses that had turned on their lights. Their dogs were still incessantly barking. It looked like a neighborhood party; people stood outside in the yards listening to all this. As we came to the end of the lane, we were met by two very large and mean-looking dogs, which meant the end of the drive for that lane. But as we stopped to turn around, the wailing was so faint, it was almost inaudible now.

Returning to the main road, we stopped at the high-tension power line cut. The breeze blew right in the truck window, and the wailing was louder than ever. Sitting there for a few minutes, I could record this onto my phone even over the engine's noise, and then it stopped. Twenty-four minutes of chaos just stopped. I sat in the truck while he turned around and returned to the house, trying to collect my thoughts. What had just happened? Back at the home, I set the recorder back up, feeling this wave of relief. Almost like a forest let out a huge exhale. Common nighttime noises started to pick back up, and I was never so happy to hear crickets! I stuck around for a while longer, waiting to hear if more knocks or vocalizations came about, but I got nothing. I was exhausted heading home.

What madness! Theorizing that a Sasquatch had just given birth, to which an entire country neighborhood had just heard, in which they most likely had no idea what they listened to anyways, and having some audio to back everything up? Madness. It was all madness, but it made enough sense to me that the next morning, armed with maps, a general location, and my forensic kits, I set out to find anything concrete to back this one up. Long story short, I looked for nearly a week and found nothing. Not even a mention in the local newspaper or online about weird noises heard on August 22nd, 2016, in Fairfield County, Ohio.

The location of the vocalizers was indeterminate at best, and I had no land access where I wanted it anyways.

Skunked again. But, what became almost immediately obvious following this night, was that the behavior of the Bigfoot changed. The audio slowed in frequency, and the aggressiveness became occasional instead of constant. Based on the events that night, it is only my theory that the prior Sasquatch mannerisms displayed towards this witness family were all attempts to protect and feed a pregnant female. Once she gave birth, the investigation simmered and leveled out. Hence, I added safety as a component to factor into this field. Clearly, not our safety, but consider what would be perceived as a threat. And in this case, I believe the witness's family was perceived as threatening.

Moving forward—

Then 2019 came around and turned out to be a very bizarre year: a rather harsh winter, no audio. The witness' property had been changing over the last year and a half, with an adjacent farmer doing serious work on his farm by dump-cutting massive perimeter trees. This led to very quick slope erosion on one finger of the hollow, causing the sandy loam soil to almost cave in and settle in on itself. This also removed much tree cover in a zone that had been notorious for great knocks. Additionally, the property had been on the decline for animals usually present. Even squirrels were only a few at best—no deer, chipmunks, raccoons, nothing. The trail cams were even void of critters. It was eerie. 2018 overall was a quiet year after a booming 2016 and 2017. I didn't have much hope for 2019. Nonetheless, it deserved due diligence. Perseverance won out because that year finally tipped the scales to a goldmine of real data. And while the audio was becoming more scarce, tracks were finally found.

2019 TRACK INVESTIGATIONS, FAIRFIELD COUNTY, OHIO

The following are my actual reports, with my investigator notes for reference. Included are photos from both investigations. A review of the methods utilized during the track line investigation is below. Pardon the shorthand.

This report is from an event on April 24th, 2019, and is as follows:

April 24th, 2019

The witness was foraging for morel mushrooms on private property in a heavily wooded riparian zone and an adjacent tract of woods. A small year-round stream flows in a north-south direction and appeared to have the influence of occasional flooding, given a wide berth. Agriculture fields are nearby on each side and above the water level of this creek. The fields had not been tilled. As the witness was looking around, they noticed an impression on the east side of the stream. The witness took the initial photo, and this investigator was notified.

The initial site investigation began at dusk. The track was observed as per the witness testimony oriented to the north. The soil was a mix of mud, leaf litter, and walnuts. Also noted was a narrow game trail, with obvious ground disturbances, leading up from the creek and opening to a top -side agriculture field. The substrate of the top field was hard earth. Using flashlights, another possible half track was noted.

Full track measured 15L x 5B x 4H

Half track measured 5 inches at the ball

Both tracks were cast.

After the casts, they were washed. Both were noted as having possible friction ridge impressions. A daytime site investigation occurred the following day.

Noticed were several small game trails and one heavily used one. Tracks observed were deer, raccoon, and pig. A small area of wetland was also noted.

This is part of an ongoing investigation.

Below are my investigator notes from the scene April 24th, 2019 with a daytime follow up investigator the following day:

Evening message. Small creek flowing from pond drainage and entering a confluence into another feeder pond to the SE. Late morel season. Behind an unplanted field inside a wide, heavily wooded tree line and wooded tract. Stormy weather. Left overnight. Substrate was rocky & soft mud. Orientation of creek bed track was heading south parallel water. Second, half track was cast along top of hillside, following a narrow game trail, at the edge of a corn field not yet plowed. Substrate was hard earth.

Full track measured 15l x 5b x4h

Half track measured 5 in at the ball

Report is from an event on July 28th, 2019 as follows:

July 28th, 2019

Trackline found on witness private property along a creek, the creek bank, and into a forested area while on 4-wheeler checking live animal traps (approximately 50ft. From waterway). The witness called this investigator at approximately 5:30 pm with report. Site investigation began within an hour. Found were 12 full tracks, one partial. In the water where obvious disturbance was noted, 3 similar impressions were noted, however not castable.

Of the 13 castable tracks:

10 were along the creek bank in a mixed substrate of pebble rocks and sand. 2 full-sized were in the mud inside the forest, and the partial was found along the short bank incline in a hard soil/ sand mix that was eventually stepped on before casting. Materials used were plaster of Paris for mud track 2 & 3 and creek track 5. Hydrostone was used for the rest. The trackline moved along the waterway in a west to east fashion (this waterway terminates in a larger body of water further east, approximately 150 yards), And abruptly turned north, up the creekbank into a long tract of forested area. It then doubled back parallel to the creek along higher ground following a heavily used game trail. 2 long scrapes were found immediately after mud track #3 heading further uphill.

Four questionable hair samples were obtained during this initial investigation: Aseptic technique was used to eliminate potential investigator contamination. All samples were placed inside paper lunch bags from an unopened pack. Labeled. Then placed inside a sealed plastic bag. Samples have been kept in temperate conditions. Photo and video documentation were obtained during the collection of evidence. Pending results from DNA analysis. Noted three lined barbed wire fence row. 2 samples from the top and second row of barbed wire fencing (1 each) 1 sample found approx. 5ft up along a sticky grapevine hanging midtrial.

1 sample obtained inside of creek track #12 where trackline originated from. Noteworthy: The two mud tracks (both right feet) were suspect for dermatological impressions along toes and other areas of footbed in ground. This proved to be an accurate assessment after bathing casts.

Average Step: 60.35 inches

Average Stride: 113.4 inches

Plentiful deer and raccoon tracks were noted in the area. PawPaw, Walnut, Oak, Cherry, and Hickory trees observed.

This area is in a unique geographical location. A freshwater spring feeds into the creek, with ponds nearby. Small recess caves and abandoned sheds are in the area. Surrounding properties are all private with farmland nearby.

This is part of an ongoing investigation.

Below are my investigator notes from the scene on July 28th, 2019, with follow-up over the next several days:

Notes:

Comparisons were visualized with the 4/24/19 left foot. Spoke with a colleague at various real times during this Investigation and asked probing questions regarding pulling off 'prints' from the three tracks with the question asked.. could this be the same suspect?

measurement chart to be attached per photo

Average Step: 60.35 inches

Average Stride: 113.4 inches

As of 7/30/19, I was phoned by witness family approximately 8 pm (who did cross over onto their neighbor's property) & reported six more possible tracks.

Pending another site investigation today for eval and follow up. ((Note: Heavy storms passed through their area yesterday (7/30) in the morning.))

#	Location	Length (in)	Ball (in)	Heel (in)
1	Woodline NE	16.25	7	6.5
2	Creekbank NE	16	7	6
3	Creekbank NE	16	7	6
4	Creekbank NE	16	7	6
5	Creekbank S	15	7	6
6	Creekbank S	16	7	6
7	Creekbank S	16	7	6
8	Creekbank S	16	7	5.75
9	IN WATER			
10	Creekbank NW	16	6.5	5.5
11	Creekbank NW	16	6	6.5
12	Creekbank NW	16	6.5	6
3*	Woodline N	16	7	5.5
2*	Woodline N	16	7.5	6

14 Tracks 13 Cast * *dermatoglyphics noted* ***hair*

Avg. Step	Avg. Stride
60.4in	111.4in
(longest 66in)	(longest 121.5in)

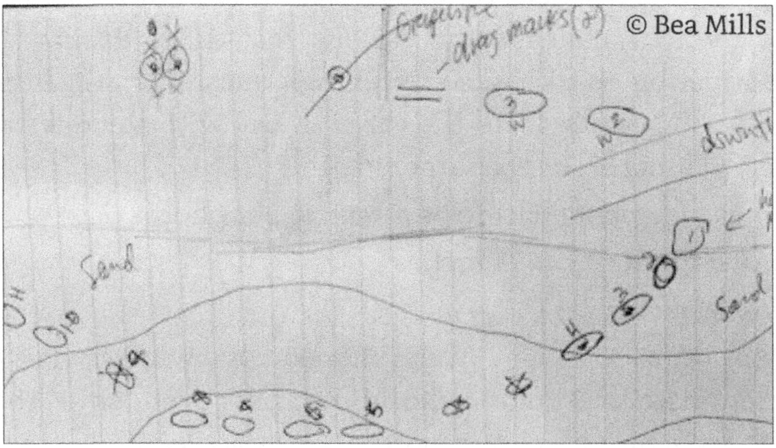

Hair collected—

Casting—

C) METHODOLOGY REVIEW

So yes, that is a lot of information there. During both investigations, the entire process was either photographed or videotaped. Each track was labeled, measured, and recorded before any plaster was poured. Over the years, I've found it much more prudent to take the time to accurately document the evidence first. This is where excitement usually overtakes common sense, and I've been in several arguments over how long I take to do this. Quite simply put, you don't get a second chance when it comes to something as extraordinary as finding a possible footprint of a beast that is not even supposed to exist, so don't screw it up. There are no redos here. Any human alterations of the initial imprint immediately negate the evidence, and you've just compromised the integrity of the entire find. If you screw something up, own it, document it as a screw-up, and carry on.

For methods related to investigating a trackline, let's focus on the second report, July 28th, 2019. While I have not heard anyone coming forth with a longer vetted trackline since, at the time of this writing, Bob Titmus and the P-G Tracks now hold second place.

Here was my method for the documentation of the trackline.

First, secure the location. Second, I made a phone call.

I already had practice casting in a sandy creekbank. Those results were okay but less than stellar. So, I called the one person I knew who had loads of experience with casting in the sand and I could trust, Cliff Barackman. Fact: The mixing ratio for plaster changes with your substrate. Too thick and you might crush your specimen, too thin and it won't cure or pick up the details you want, etc. Mixing plaster is somewhat of an art. Even different plasters mix differently.

Hydrostone mixed too fast will leave air bubbles, so you must let it settle first without hardening. Plaster of Paris is nearly the opposite. Practicing with your materials is something I recommend to everyone, even just for fun! You can practice outside with the animals, your feet, mud, sand, and snow. Check out your local parks for their programs, join me on a BeaOutdoors Class, or come to the Hocking Hills Bigfoot Conference—we all have track casting programs. The point is, practice! It is the only way you can be more prepared for success should the occasion arise that you want to cast something while in the field.

Join me at the Hocking Hills Bigfoot Conference and BeaOutdoors Classes!

In this case, I wanted someone I considered to have expertise with casting on this type of earth. He was spot on helping me with the plaster mix recipe on the more complicated tracks. Thank you, Cliff. I keep a baggie of bamboo skewers with neon-colored notecards and ribbon with a sharpie in my to-go tote. This has proven more than once to be worth its weight in gold. It photographs better and helps alert others (and yourself) to locations of interest. Of course, the tape measurer is always in my to-go tote. I had video and cell phone cameras on this site investigation.

I stood along the opposing creekbank with the witness where he first saw them as he explained the events. Raccoons were bad in the area, and one had recently torn up his wife's vehicle causing several thousand dollars worth of damage. He had been setting live traps to catch these animals, and one was positioned along the upper portion of the creek. This particular evening, he rode down to check the trap, and when he looked along the creek. Seeing questionable tracks along the bank, he called me and sent photos. He had not interrupted the site to investigate closer, and I saw no nearby or adjacent human disturbances upon my arrival.

After securing the location, I took general, wide-shot photos and notes. I pulled notecards and a notebook from my backpack. We slowly and deliberately walked to the track nearest us. Each track was numerically labeled. At least three photos were taken of each track found with the tape measure indicating the length and width at both the heel and ball. These were written down twice to prevent errors—once on the numbered notecard (and photographed with the data) and once in my field book. It sounds ridiculous, but when you're looking at hours' worth of work, do it right the first time! The creekbank was not ideal for sticking the skewer into; I just laid the card next to each one.

After each track, the respective step and stride were measured. There is a method for measuring each, but if you remember nothing else, remember this - always measure from the heel, never the toe. Why? Because toes have joints and move about, toes splay. The heel is fixed and, while it can widen, does not shorten. That said, measuring a step is merely one heel to the next heel, one footstep to the next footstep. (eg. left foot to right foot, and visa versa) On the other hand, measuring a stride is to measure two actual steps—left foot to left foot or right foot to right foot. Each track's step and stride were measured, even the #9 in the middle of the creek. Yes, this took hours. Daylight was becoming an issue along with the threat of rain coming, yet it proved invaluable for the records. We left the creekbank in search of the rest of the trackline. Following the obvious step direction, we went up the opposite bank and into the forested area. This was the side of the creek that the tracks entered from, so in short - it doubled back. The land corridor is typical for a creekside, so the immediate area was now littered with thick tall grass, multiflora rosa, stinging nettle, and a large felled tree, nothing pretty when looking for tracks. Now, employing the slow and methodical process where using those Step and Stride measurements from earlier, with a little orienteering knowledge, becomes useful and absolutely necessary.

Thankfully the witness is a tracker, so this went quickly, and we could track the steps. They eventually led us to a well-used game trail that ran parallel to the creek and further up the ridge. We had made an almost complete circle from the entry point at the stream, using a tree for reference. Fortunately with the recent rains, there was a long swath of mud along the trail, and two more tracks preceded by a long, narrow grooved slide were found. These tracks were pristine, documented, and cast.

Upon gross inspection, I was curious if the little lines in the mud were possibly dermatoglyphics (fingerprints, but on the toes.) I documented everything before casting, and since there were no second chances for casting here, it was horrifying. And yes, it was also a perfect pour.

Also, to note. At the entry point of the trackway to the creek, three sets of deer tracks were found; only two sets of tracks made it out across the bank. Track numbered twelve (the entry track) had obvious deer hair inside the footbed. An aseptic technique was used in hair collection before the mix was poured. I know the average person with a microscope at home can identify deer hair without question. But curiosity compelled me on this occasion in that, since deer are easily identifiable and this mess of hair was smack in the compressed part of the footbed, could it be possible that 'something else' might present itself in a DNA analysis? I had enough hope and reasonable suspicion that it was worth a shot. Three other questionable samples were collected along the line we tracked. My evidence collection was videotaped for quality. All specimens were labeled and marked for the chain of custody, then readied to be sent for analysis. Of course, using forensics in specimen collection is a whole other chapter in itself, although the same rules apply - you only get one chance to not screw up. And by screwing up, I mean contaminating the entire specimen in one breath, one lazy finger, sealing the bag wrong, your hair being unkempt, sneezing, etc., aseptic means aseptic. When speaking of wanting a DNA specimen that traditionally comes back as human and hoping for something else, you had better make sure that it's not your own DNA mixed in there, or the joke is on you and whoever wasted their time on your poor collection. This is a heavily abbreviated version of the entire event. An exponential amount of details were omitted.

In summary, the active investigation continued until almost three o'clock in the morning, pouring plaster, praying for no rain, finishing notes and sketches and I had to be at work in a few hours. I continued to revisit the area for the next several days and found two more tracks of the identical measurement. And yes, the recorder was on the entire time. Case Study Complete. In follow-up, these dermal ridge suspect casts were personally taken to be examined by both Dr. Meldrum and Cliff Barackman. I felt it was pertinent also to include the April tracks report because they were used in comparison while assessing the July trackline to determine eligibility for repeating patterns. The overall conclusion was positive: I had not just wasted the last four years of my life. I cried.

I had trusted another investigator to send the hair samples to a supposed colleague for DNA analysis. After waiting for months and being told the entire lab team was sick with the flu in South America, I was informed that the samples had come back as raccoon hair during a face-to-face conversation. Except one, and it was a plant fiber. I already knew two definitely had deer hair in the samples; the plant fiber was a long shot. However, from all the specimens, I was hoping for that 'something else.' Any way the microscope turns, they were not raccoons. I updated the witness family about the results, and from there on, all investigators were banned from the property, and I was not to release any evidence without prior discussion. My hair samples were gone. The chain of custody was broken. So, I will never have true data with regard to the samples submitted. I regret not saving a control. A huge loss. This case study, with supporting field work, additional casts and documentation, was the actual investigation that awarded me the '*Bigfoot Times* 2019 Bigfooter of the Year.' First Female.

Things to remember: Always keep a control specimen. Document everything. Take your time. Give due diligence and, most importantly, be careful who you trust.

Chapter 5

PRINCIPLE #4: Resolution

"If You Have No Expectations, You Will Never Be Disappointed"
-Hansen Fong, Professional Photographer, Author, Friend

This is the last part of the puzzle. We've covered pieces, strategy, and Execution and have reviewed a real investigation case study. Now we're left with choices. What to do with all the data gathered? Social media is a razor's edge. The old saying, believe only half of what you read, a quarter of what you hear, blah, I think, is even more diluted now, compliments of the internet. Anyone can put on a smile and lie through their teeth. Hoaxers, thieves, and storytellers further pollute the airwaves. Putting this data into a video with clickbait, dramatic music, and long introductions will never be the answer. If anyone sends me something to watch, the more dramatic the intro music, the faster I turn it off. Posting this data on a crack book page with no context is ridiculous, eaten alive, and rightly so. The burden of proof rests on the presenter. I have no intentions of using my work for 'clicks' or 'popularity' or whatever other madness. I fell into this field because something strange happened, and I wanted the truth. I wanted answers. And by being in the role of an investigator now, I wanted to do my best to document as much as possible for the future of this field and improve on my own techniques. I don't know everything, I never will, but I love to learn, discern and use those bits of information to better my own work. No one wants to take up a hobby and never get better, do they? I would venture that the answer is -No. People take up hobbies to learn new things and skills and have fun.

But what happens when you find yourself in uncharted territory, and the help you need is deemed so insane to people not in this field that you just get scoffed at, or worse? So really, what DO you DO? And, WHO do you trust?

Academia? -You stand alone.

Call an expert? -Who exactly would that be?

Is there Logic to any of this? -Go Fish!

What are my next steps? -Choose your own adventure and carry on.

You have only your wits and common sense when adopting the principles of a citizen scientist to guide you. At least, that's what I did. I muddled through years' worth of data until I decided what I wanted to know. And to this day, I still have so many questions and so much work to

do. And I wholeheartedly believe I'm still missing clues hiding in plain sight, whether through my notebooks, scribbles, photos, videos, or even the casts themselves. This is where getting your head out of the box is most important. The truth is, many people with wonderful academic backgrounds do hold answers to most of the questions; it's just your presentation of THE question and being specific in your query. Not all your paths will lead to the answers you want. Sometimes, it goes in weird directions. And sometimes, you just scrap the entire question and return to the beginning. I will repeat Mr. Dick Geyer here, knowledge for the sake of knowledge.

I attended a wildlife conference a few years back, wanting to hear the presentation of a trophy whitetail deer hunter. I look to the hunters and farmers when I have specific animal behavior questions. I always wondered why we buy Flir's instead of rifle scopes. What's up with Scentblocker? Tell me why humidity is a thing. After listening to his presentation, I had even more questions.

I was stuck on finding a pattern during this investigation. I was certain there had to be a pattern, and I was missing it. I went up to the gentleman after his talk and specifically asked him if he noticed that the deer had patterns from year to year. He looked at me quizzically, yet I implored with, do you notice the same buck or herd frequenting the same ground year after year? His answer was so wide I just thanked him and walked away. He said no, not really. I didn't even know what to do with that information. It made me sad after watching several videos of hunters who, year after year, watch a buck grow up. It reminded me that those were just videos for sale or clicks, and this fellow was a person who had no reason to lie to me. Simple and sad, but I was still unconvinced.

Having some time off work, I used it on my hobby. Sitting down with an entire brick of coffee, I spent nearly a week working to pull every stitch of information (and kicking myself for not doing it in real-time) on my audio files. Figuring I had thousands of hours and multitudes of audio clips over a four-year period, I got to work looking for the thread. The time and date stamps were already recorded. But what details had I overlooked? So, I got on two different websites to write down every detail I thought was pertinent. But what about the moon phases, barometric pressure, precipitation, time of sunset, time of sunrise, or relative humidity—? It was an exhaustive list. I wrote every one down for every single vocalizer audio clip I had. If there were multiples in one evening, I simply stayed on the date and did not count them individually.

My goal was to look for anything that could be referenced as a pattern. With some humor, I sought to see about the old full-moon debate. The East Coast vs. West Coast Bigfoot! The theory goes West coast Bigfoot are active during the full moon, and East coast Bigfoot go silent until the full moon is over. Having four full years with 12 months and 13 moons—that's a lot of data, and I wanted to see if I had an answer. You can see for yourself how that data turned out based on the measurable parameters.

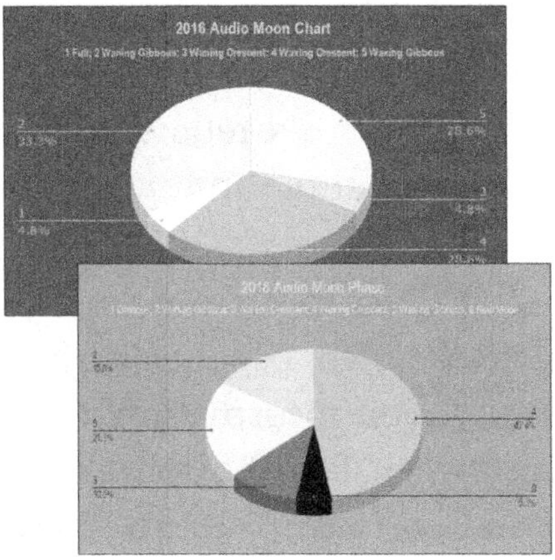

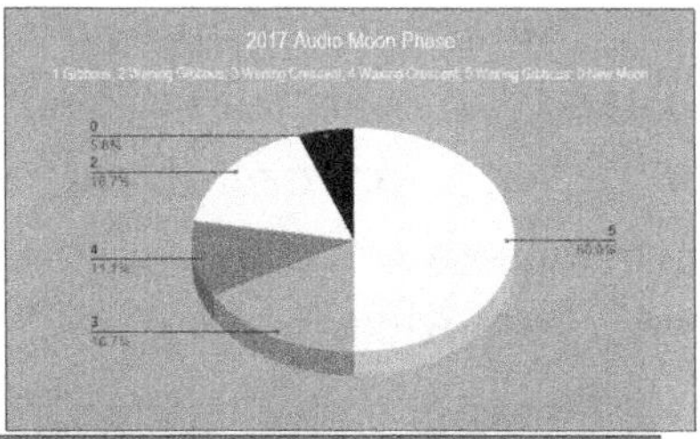

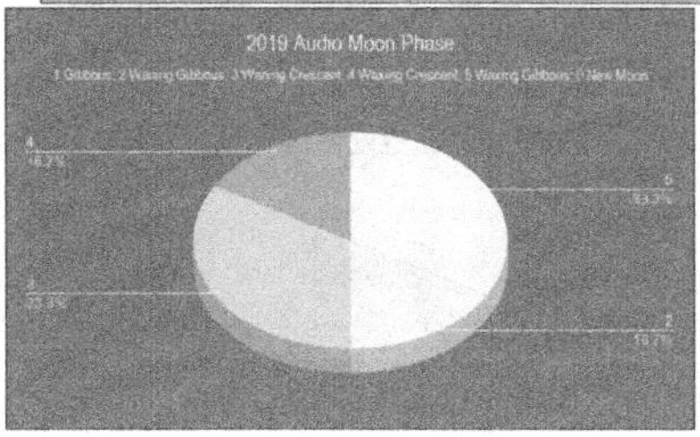

BUNK! By Ohio standards, at least. I cannot, in good faith, say that the East coast theory is right. (2016 had no vocalizations during the new moon, either) The idea started out funny, but I was sad to see my results. The waxing and waning gibbous phases of the full moon were usually sure-fire vocalizer nights, weather and season dependent. But even with all my info, I couldn't put the joke to rest. So, that theory will continue for my eternity. I put together all sorts of graphs and charts based on the data, and not one came back with anything I could point to or put my finger 'on paper' as an 'Ah-Ha, I found it' moment. I still have this burning in the back of my gut that I'm just not looking at the data right. Time will tell, and maybe a fresh set of eyes or a glimmer of new knowledge will reveal something.

But who am I anyways? This is just one point of data, and probably a million other things can be done with it. So, while the jury may forever be out, I had one focus: looking for a repeating pattern hiding inside the audio data. And so far, the answer is—there is none.

What else? Based on the very educational conversation between Cliff Barackman and Dr. Meldrum while inspecting my casts in 2019, one question kept coming up. It was asked if the dermatoglyphics were repeated in all the casts, meaning, could the three different locations, during three different times be the same individual? That proved much harder to determine than I originally anticipated. Now dermatoglyphs are the impressions on your digits. Most people are familiar with 'fingerprints.' Our toes have the same thing, and we do call them toe prints. These are found on footprints and are very individual to us all. Fun fact, the koala has fingerprints that are nearly identical to ours!

I started a Facebook page with hope. Hopes of generating ideas and going through the entire process of documenting and examining the toeprints in question with other colleagues - from start to finish. I was hoping this would be an opportunity to finally open the scientific community's eyes to establishing this fairytale 'Bigfoot' beast as a species.

According to a friend, he remarked that the late Grover Krantz had suggested this: repeatable prints may be enough to prove the species exists, and I held that close to my heart. If I could somehow use what evidence I had documented and collected, it might help prove the existence of Bigfoot. I thought that being able to present hard data would be fantastic. And lastly, I hoped, through the documentation, that if someone else had the good fortune of casting a great track, my work could serve as a decent guide to follow. Why reinvent the wheel? Learn from others' mistakes. And boy, did I nail that last one.

The first attempt to determine if the toeprints were repeatable was a glorious success by a skilled craftsman making latex molds directly from the casts. They were beautiful. I later sought more advice, trying to figure out what I could fit inside the mold to maybe ink, dust, or press out. I was told to use glue, so I did and poured the soft material into the latex. Horrified, they ended up adhering to themselves irreparably.

Yes, I ruined these molds and had no second chances. I was terrified to try again because I did not want to compromise the integrity of the impressions on the cast. I just stared at my ruined molds and prayed for better technology. This was over until I could figure out a better way to pull forensic data without destroying the casts.

And while technology does keep improving, I had no idea where to turn. I had to be fair to the scientific process of trial and error when in truth, I was mad. But true madness was closer than I could ever imagine.

Life suddenly got strange and bizarre. Just after being awarded the *Bigfoot Times*, Bigfooter of the Year at the end of 2019, the fresh, new 2020 year began with the global outbreak of Covid-19. Like a horror movie from afar, I quickly realized I had no idea what horror meant. Among many, many other atrocities, I was sick for nearly six weeks in early March. I withdrew from co-hosting the 2020 BFRO expedition not just for my own health and safety but for the protection of future campers. I had now been cast back into the front lines of COVID, working the night shift in the emergency room while people and patients were dropping like mayflies around me. Standing amid a Tsunami of horror and madness, I decided to scrap the entire idea of the forensics Facebook page and stop pondering hobbyist questions, and prioritize that my life trumped Bigfoot. The answer to life, the universe, and everything, is not 42. It's DOCUMENTATION.

Despite my best efforts to care about these hairy folks of unknown origin, I felt like, who cares? Would proving the species take away from the fun and the mystery? Would it ruin product sales or make them better? Conferences and festivals, merchandise galore? Would it destroy the entire belief systems of some folks? Would it make others happy to put its existence to rest finally? I have no idea. It's hard to even speculate on such things. Bigfoot is supposed to be fun. It's a hobby. What I do know is that after five years of this investigation, I've found myself strangely overjoyed for urban sprawl. I am confident that the Bigfoot that prowled the property for decades have moved on. FOR SALE signs are on the nearby fields; I suppose it was just a matter of time.

Nearly every week, new homes are popping up. Mega Industry moved in next door. I never thought I'd be so happy to say this investigation is closed. In late 2021, I was utterly relieved and overjoyed when I picked up vocalizations in a completely different location. Using my four principles and a bit of that 'gut instinct,' it was somewhat emotional hearing them again. The old ways and old days are long gone so I guess, like everything else, the best thing to do is move on.

Madness.

Chapter 6

Now Here

Why on earth get involved in the search for Bigfoot? Why wonder day after day just how that mythological being is doing? Daydreaming of the perfect daytime sighting. Planning trips all over the country, chasing things that go bump in the night. After all this? WHY? I will tell you why.

Because Bigfoot is a beautiful reminder that everything impossible is possible. Despite anyone's best efforts, there is insufficient proof that such a being exists. I failed. And even if I hadn't, it most likely would have ruined the mystery for many. When I had my first sighting in 2014, Bigfoot reigned in my curiosity, and the fun was gone. It all turned into seriousness in that quest for knowledge. If curiosity really did kill the cat, I don't know how I am still here. Miracles, perhaps?

So again, WHY! Why am I still doing this? Easy. Because in the quiet of darkness, I remember why I started in the first place, it was fun. It was just plain old, nature-nerd fun.

Remembering the old days camping as a girl scout and bringing back those old skills in handy times. Chatting with others about new gear and implements. I have traveled to the most remote of places, visiting little towns, eating in their diners, reading their newspapers, and enjoying life. I've gotten to hike in places normal people don't go. I've seen so many beautiful things going on a simple walk. I've explored areas that I would never have gotten to see—absolute gems of nature with their own unique landscape, weather, plants, and critters. I've met so many folks from all over the world. I heard amazing stories and even more fantastical ones. I've learned how different folks look for Bigfoot and exchanged ideas and theories—from intentional

settings, quiet prayers, and a very holistic feel to almost militant and exacting with precise coordination of plans and everything in between. I've learned much just from being present, patient, and quiet with other folks.

Styles even differ based on region. It's a blast! The campfires were always nice, sharing a warm glow and keeping summer humidity down and winter temperatures up. Oh, the conversations that could go on for hours and pick up days later. Insights and friendly debates over all sorts of life topics. Politeness, kindness, and this semblance of kindred-spiritedness are so refreshing. During my adventures, I love taking photos and sometimes posting them online.

With respect to an old neighbor and witness, Ruth Fink—for several years, she would send me private messages thanking me for sharing those photos since she couldn't leave her house. She inspired me. If even one person got to smile seeing my silly photos, that was a good enough reason to continue. Rest in peace, Ruth. I hope you are hanging with the Squatches laughing at us down here now; seeing the countryside and learning backcountry history is one of my favorite parts of 'Bigfooting.' Getting to visit a witness on location, watching and listening to them retell the event and help look for clues or answers, it's just fun. All the new things to learn about different areas. Fun. Bigfoot. Is. fun. Some folks even look for Bigfoot without ever leaving their vehicle! So yes, literally anyone can do it!

I adore getting the kids involved; they usually ask the best questions anyways. They still see nature with awe and wonder. The year I started the Hocking Hills Bigfoot Conference in 2016 at Camp Oty'Okwa,

I was touring the grounds during a camp session. The woman I was with stopped and found a tiny green inchworm. She handed it to this little boy with a recycled milk carton slung across his shoulder with a tie used as a water bottle and rested it on his finger. The look on that young child's face is one I will never forget because it broke my heart. His eyes went wide, and he just stood there frozen, looking at this tiny green mystery inch its way along his finger. He asked what it was, and she smiled, answering him. I felt a calling that Bigfoot could be used to help kids experience things like this—nature.

After seeing that one inchworm and that one face, I immediately made a promise, The Hocking Hills Bigfoot Conference will continue as long as I live and breathe for those kids. This event has presented over thirty thousand dollars to the camp, and no child pays. The counselors are amazing with the biggest hearts. The Big Brothers/ Big Sisters Organization is worth my time for all they do. From Crisis camps to grief camps, day camps, school programs, resident camps, and everything in between, this is just one reason I do this. I was also inspired to reach out to the Hocking Hills Region and Appalachia, using Bigfoot as a catalyst. I learned about the Logan Theater's project to restore and reopen those old doors from days past. I also learned about the Hocking Hills Children's Museum and how they endeavored to help the kids as well. I felt I had an opportunity to help them, promote nature, educate, and help strengthen the communities, small business owners, and tourism. I was interviewed in the Logan Daily Newspaper in late 2021, and decided that Bigfoot finally needed a home, so why not, Logan? 2022 was my first year hosting the Inaugural Hocking Hills Bigfoot Festival in downtown Logan, Ohio. The list of thank you's to the folks who helped put this together is colorful and long, but this town and I saw over 11,500 people come from across the United States.

During the event, I was awarded an accolade from the State of Ohio House of Representatives for helping the region. Of course, I cried. It was unexpected. I was humbled, and it was epic. I look forward to many more years filled with fun, families, smiles, and good causes. Fun is the key, and Bigfoot is to blame.

In 2019 I started my organization, BeaOutdoors. It's taken a few years, but I now have several programs and classes specifically designed to elaborate on my Four Principles. Now in full swing with certification. My goal here is to help folks learn and keep it relatable for use in their work and life.

I'm always looking for more opportunities. I have searched for ways that the mystery of Bigfoot can continue to be fun and educational and get kids, adults, and families involved while investing in and giving back to the beautiful Appalachian region. I feel true satisfaction in helping others explore the outdoors and welcome anyone to join me.

Whether it is coming on one of my BeaOutdoors expeditions, attending one of my conferences, or simply wanting to get active in your area. Check it out, and see what sparks your interest. Critters are cool. Nature can be scary. And there is so much to learn!

Most importantly, now, I do this for myself and my dad so that he can tell stories of my nutty adventures to his friends in town. For my nephews. For my mom, a decorated Navy Captain, Ret., one of the first women to be a flight surgeon, breaking glass ceilings for women in the 1970s and me in this fringe field 40 years later. I do this for those who want to know things, be it invasive plants, trees, land navigation, ecology, nature in general, or even Bigfoot. I do this to help the kids who might otherwise never get an opportunity to experience the outdoors. To promote a culture of caring and openness and choose to leave a legacy for all outdoorsy folks. I do this to share beautiful pictures for people who can't get outside.

I do this to continue my quest for knowledge and share what I know. I do this to capture memories, to look back, smiling that I finally saw a scorpion wasp chilling on my hand. I do this for my friends and co-workers, who enjoy a good laugh on Monday mornings at my antics. I do this for Suzanne, my mental floss. I do this for my son so that he may never forget that it's okay to do things your weird way and remind him that monsters do exist. I do this for women because I can. I do this to raise the bar on my field methodology. Warranting that the -ology part means "the study of"—therefore mutable and able to change based on advancing technology and field trials.

And I do this for you as both a hopeful learning tool and a cautionary tale. I hope that through my own experiences, you can become even better investigators and enjoy this hobby even more. Remember and Heed my warning.

There are exactly zero Bigfoot experts. Are my principles the end all - be all? Absolutely not. We are as unique as the day is long. But, I hope that some of the methods here can serve as a guide for you to build your own. Bigfoot is a mystery. Bigfooting can quickly become madness. The truth sometimes sucks, but the truth is what sets us, and hopefully that hairy guy someday, free. Knowledge for the sake of knowledge. Mr. Geyer's legacy will carry on. Maybe even with a little help from Bigfoot.

I hope your adventures take you to great places. Just remember, anywhere life does take you, you will always be - Now Here.

End.

To Babayka, Thank you

Made in the USA
Monee, IL
04 November 2026

67385302R00056